Henryk Matusiewicz

Potential release of in vivo trace metals from metallic medical implants in the human body

AF154040

Henryk Matusiewicz

Potential release of in vivo trace metals from metallic medical implants in the human body

From ions to nanoparticles

LAP LAMBERT Academic Publishing

Impressum / Imprint
Bibliografische Information der Deutschen Nationalbibliothek: Die Deutsche Nationalbibliothek verzeichnet diese Publikation in der Deutschen Nationalbibliografie; detaillierte bibliografische Daten sind im Internet über http://dnb.d-nb.de abrufbar.
Alle in diesem Buch genannten Marken und Produktnamen unterliegen warenzeichen-, marken- oder patentrechtlichem Schutz bzw. sind Warenzeichen oder eingetragene Warenzeichen der jeweiligen Inhaber. Die Wiedergabe von Marken, Produktnamen, Gebrauchsnamen, Handelsnamen, Warenbezeichnungen u.s.w. in diesem Werk berechtigt auch ohne besondere Kennzeichnung nicht zu der Annahme, dass solche Namen im Sinne der Warenzeichen- und Markenschutzgesetzgebung als frei zu betrachten wären und daher von jedermann benutzt werden dürften.

Bibliographic information published by the Deutsche Nationalbibliothek: The Deutsche Nationalbibliothek lists this publication in the Deutsche Nationalbibliografie; detailed bibliographic data are available in the Internet at http://dnb.d-nb.de.
Any brand names and product names mentioned in this book are subject to trademark, brand or patent protection and are trademarks or registered trademarks of their respective holders. The use of brand names, product names, common names, trade names, product descriptions etc. even without a particular marking in this works is in no way to be construed to mean that such names may be regarded as unrestricted in respect of trademark and brand protection legislation and could thus be used by anyone.

Coverbild / Cover image: www.ingimage.com

Verlag / Publisher:
LAP LAMBERT Academic Publishing
ist ein Imprint der / is a trademark of
OmniScriptum GmbH & Co. KG
Heinrich-Böcking-Str. 6-8, 66121 Saarbrücken, Deutschland / Germany
Email: info@lap-publishing.com

Herstellung: siehe letzte Seite /
Printed at: see last page
ISBN: 978-3-659-56488-8

Copyright © 2014 OmniScriptum GmbH & Co. KG
Alle Rechte vorbehalten. / All rights reserved. Saarbrücken 2014

Table of contents

1. Introduction

Over the past decades, there have been significant developments in the availability of suitable metals and metal alloys for orthopedic and dental implants. These materials possess excellent corrosion resistance, good mechanical properties, low inherent toxicity and good biocompatibility. However, when pure metal and metal alloys are implanted into a complicated and corrosive physiological *in vivo* environment, the surface oxide film stability may be affected (disrupted) and fresh metal surface must be exposed to release a large amount of metal ions from metallic materials. resulting in increased metal ion release. Metal ions are released to the aggressive body environment (biological fluids and tissues in the body) from metallic biomaterials, used in such procedures as total artificial joint arthroplasty, instrumental arthrodesis of spinal segments, bone plates, screws, dental implants, *etc*. This is accomplished through various mechanisms, including corrosion, wear and mechanically-accelerated electrochemical processes, such as resulting in stress corrosion, corrosion fatigue and fretting corrosion. Released metal ions may cause various deleterious phenomena, inciting allergies ("metal allergy") and potentially promoting granuloma formation and even carcinoma. A large amount of released metal ions could be harmful to human health and may eventually lead to severe complications and failure of the implant system. The mechanisms of metal ion release from metallic materials *in vivo* need to be understood in order to discuss the safety and biocompatibility of implant materials. Since the release of metal ions depends on electrochemical rules, much effort has been made to analyze the electrochemical processes and investigate metal ion release. As an implant corrodes, either electrochemically or mechanically, metal ions, metal complexes or implant-derived particles (in the nanometer range) are released.

Medical grade stainless steels and other alloys (Cr-Ni; Co-Cr-Mo; Co-Cr; Co-Ni), tantalum, zirconium, wolfram, titanium and titanium alloys, magnesium and its alloys are all used as biomaterials, while noble metal-based alloys are used for dental materials because of their high corrosion resistance and durability. Metal ion release from metallic implant is inhibited by the surface oxide as a passive film where partial dissolution and reprecipitation are repeated in aqueous solutions, therefore ensures corrosion resistance. Therefore, the film properties and the potential to change *in vivo* must be considered in order to understand metal ion release. Additionally, being able to quantify the released ions is important and the behavior of the released ions and their potential toxicity must be studied and understood, because they may not always have a detrimental effect influence on the human body.

2

This paper takes a close look at the mechanisms of metal ion release from metallic implant materials and the behavior of the released ions *in vivo,* presented according to empirical data. This review will help in the establishment of an analytical methodology for the accurate and precise determination of the main components of metals and metallic alloys used in articular prostheses and dental implants. After a brief summary of the composition and types of bioimplant materials and biocorrosion, the release of *in vivo* metal ions is discussed. Other aspects leading to the release of *in vitro* metal ions and potential related adverse physiological effects, including toxicity, carcinogenicity, genotoxicity and metal allergy, are beyond the scope of this contribution and will not be discussed here. In accessible literature the majority of papers have established an *in vitro* approach which is not reliable to *in vivo* and/or *ex vivo* conditions. It is not within the scope of this review to evaluate possible toxicological effects produced by the intake of trace metals, nor to estimate tolerable body intakes according to the recommendations of health agencies.

For the first time, a comprehensive survey of available approaches for the determination of *in vivo* bioaccessible metallic nanoparticles/ions in the human body is reported, including a description of commonly used sample preparation procedures and measurement techniques used for analysis of derived biological (clinical) fluid and tissue samples. The present report was designed to investigate the alteration of body fluid and tissue levels of trace metals in implant patients, as compared with controls, by performing a systematic analytical review in papers which report the state-of-the-art in the research and generalize results of research on exposure and bioavailability of metals to human in *in vivo* tests, using different techniques of metal analysis. Furthermore a compilation of the results (metal ions studies) that have so far been published in the literature is presented (up to June 2014).

2. Biocorrosion and wear behavior of metallic medical implants

Chemically speaking, corrosion is the visible destruction of a metal. Electrochemical reactions during which the surface of a metal is deteriorated *via* ion release are called *corrosion*. Corrosion of metals and alloys used as implants in the body is a complex process that the metal is challenged within the body due to changes of the pH, body fluids, exposure to cellular processes, *etc.* and is due to the chemical environment of the body [1-3]. No metal or metal alloy is completely inert *in vivo* [4]. A metallic "foreign body", such as metal-on-metal bearings or dental implants, may interfere with oxidation-reduction reactions that are the basis of metabolic and growth processes. Redox reactions occur at the metal surface and can

3

cause denaturation of the tissue that is in contact with metallic implants. The metals and alloys used as surgical implants achieve passivity by the presence of a protective surface oxide film (layer). This oxide film, called a passive film, inhibits corrosion and keeps current flow and the release of corrosion products at a very low level [5]; the film determines *in vivo* metal ion release [3]. Even so, metallic ions from orthopedic and dental metallic implant materials are eventually generated and released into the body by electrochemical corrosion of metal surfaces, chemical dissolution, *in vivo* wear, or mechanically accelerated electrochemical processes, such as fretting corrosion, stress corrosion or corrosion fatigue [5a,6-9]. However, less attention has been focused on particles generated by corrosion, perhaps because evidence of macroscopic corrosion in the current generation of single-part components is rare.

Nanoparticles can be introduced into the human body by ambient pollution, for example, nanoparticles originating from combustion, diesel exhaust and so forth introduced through the respiratory system, or nanoparticles from sprays, creams, sun lotions, *etc.*, introduced through the skin. The risk from nanoparticles will depend on their size, structure and chemical composition. The fact that nanoparticles can originate from various devices implanted in the human body is very well known in orthopedics; the question is what is the limit to the level of nanoparticles that is still tolerable in the human body. The presence of metallic particles in peri-implant tissues may be due, not only to electrochemical corrosion, but also to frictional wear or a synergistic combination of the two [10-13]. Numerous articulation cycles produce billions of metal micro- and nano-sized particles *in vivo* [14,15]. These metallic wear particles are accumulated in the surrounding tissues or transported by synovial fluid, blood or urine to distant organs. Free or phagocytized metal particles are transferred by the lymphatic system and their accumulation in the regional lymph nodes, liver and spleen have been confirmed in postmortem studies [16]. The degradation of metallic implants induces the transfer of some metallic elements from prosthesis to the surrounding soft tissues [17] and to the liver and kidney [17a]. Doorn *et al.* [18] reported on a transmission electron microscopic analysis of metal particulate debris retrieved from 13 patients having revision of a MoM total hip replacements (THR). The majority of the Co-alloy wear particles were less than 50 nm (range, 6 – 834 nm), approximately one order of magnitude smaller than what has been reported for retrieved polyethylene (PE) wear particles. Metal wear particles from metal-on-metal THR were present within macrophages, singly as well as in apparent agglomerations [19-22]. The best method that has to be considered in order to reduce the metal ions

4

levels is to reduce the bearing's wear debris production. In general, the generated metal particles (wear debris) sizes are within the range between 40 – 50 nm [23,24]. Metallic debris may also be distributed throughout the vascular system as ions or particles [25]. Passi *et al.* [26] reported diffusion of titanium into the bone bordering dental implants, while when aluminum was present in the fixture, it leaked diffusely into the surrounding bone. Vanadium leakage was not found in the tissues.

These events cause release of microparticles of metal, which can become integrated into the soft tissue surrounding the implant. The metal particles also undergo corrosion, resulting in metal ions entering and circulating in blood. However, modern devices, made in accordance with sound metallurgical practice, using "clean" material and fabrication processes, rarely if ever show visible signs of corrosion.

3. Metallic/bioimplant materials used for medical applications

Biomaterials are either natural or man-made materials, which are used to aid or replace the functions of living tissues. They must simultaneously meet many requirements and have special properties, such as non-toxicity, corrosion resistance, fatigue durability and biocompatibility. Metal and metal alloy biomaterials (stainless steel, Co-, Cr-, Ti and Ti-based alloys) for implants have been widely applied in the orthopedic and dental fields. These materials possess excellent corrosion resistance, good mechanical properties and biocompatibility. However, when a metal or metal alloy is implanted into a complicated and corrosive physiological environment, the oxide stability may be affected, resulting in increasing metal ion release. Therefore, the analysis of impurities in steels and alloys used for medical applications is of prominent importance, because of possible health effects from the use of improper materials. Arthroplasties are used for the correction of damage to the human skeleton and for dentures and fillings. For these applications, highly corrosion-resistant alloys are used. Another important field for the application of metals in medicine is the use of steel skin staples for wound closing. Materials in direct contact with the human skin or body fluids must be selected with great care. They must be highly resistant to the corrosive effects of body fluids in order to avoid biological reactions causing deterioration or abraded particles. Furthermore, they have to be workable. Unsuitable materials or even the smallest impurities in the steels or alloys used can lead to allergic reactions or delayed wound healing. Therefore, analysis techniques used for these materials should allow for the determination of main, trace and ultra trace element amounts.

Stainless steel, C-based alloys, zirconium alloys, tantalum alloys, niobium alloys, pure titanium, Ti-based alloys and biodegradable alloys (Mg alloys and iron-based alloys) are all used in various orthopedic implants [27-31], while noble metal-based alloys are used for dental materials, because of their high corrosion resistance and durability [32].

In medicine, the most often used form of stainless steel has the denotations AISI 316L (surgical or orthopedic stainless steel), ASTM F-55 and F-138 and contains 17-20% Cr, 13-15% Ni, 2-3% Mo, as well as small amounts of other elements [1] and is the only biocompatible stainless steel. However, its wear resistance is poor in comparison with other metal implants [33]. The notation "L" indicates that the steel has a low carbon content (<0.03%) and is therefore not susceptible to intergranular corrosion due to precipitation of Cr-carbides at the grain boundaries [34]. The corrosion resistance of stainless steel is based on the formation of a thin passivation layer, mainly containing Cr (III) oxide (Cr_2O_3). Stainless steel implants are used as temporary implants to help bone healing, as well as fixed implants, such as artificial joints. Use of steel as permanent implants such as in joint replacements has decreased since Co- and Ti-based materials have become available. However, steel joints are still very popular and have an appreciable market share. Apart from the lower corrosion resistance the possible development of cutaneous metal sensitivity to Ni or Cr in patients receiving stainless-steel-based implants has also restricted its use [35]. Basically the allergies have been documented and are clinically shown [36]. Normally the body contains approximately 3-4 g iron. Iron is enriched in the surface oxide film and nickel, molybdenum and manganese are enriched in the alloy substrate, just under the surface oxide film. Clinical experience and accumulated knowledge suggest that the human body tolerates leachables from steel relatively well. Nevertheless, efforts have been taken to develop new, Ni-free stainless steels.

The major alloys in orthopedic implants are described by the American Society of Testing and Materials (ASTM) [34,36] by the International Organization for Standardization, published by British Standards [37] and by ISO 5832-6 [38]. Cobalt and chromium are the major constituents of these alloys and, therefore, is of major interest. Cobalt-chromium alloys contain a cobalt base, in amounts of 66-67% with an additive of 26-30% chromium, which increases the hardness and corrosion resistance to form a passive layer on the product surface. In addition, these alloys contain nickel in amounts of 3-5%, which increases plasticity and malleability. Furthermore, the addition of molybdenum in amounts of 4-5%, and other additives in amounts of less than 1% yields fine-grained alloys and improves the technological properties. Cobalt-chromium alloys can be described as non-magnetic, wear- and corrosion-resistant and stable at elevated temperature. The surface oxide film of a Co-Cr-Mo alloy contains oxides of cobalt and chromium without molybdenum. The corrosion resistance of Co-based alloys, similar to stainless steel, is based on the formation of a thin passivation layer of Cr_2O_3.

Titanium (CP-Ti, ASTM F-67) and titanium-based alloys (Ti-6Al-4V, ASTM F-136-02a and Ti-6Al-7Nb, ASTM F-1295-05, Ti-15Zr-4Nb-4Ta, Ti-40Nb) have been widely used in orthopedic and dental implants and can be considered as the most corrosion-resistant of the alloys described. This is based on the very high stability of the TiO_2 passive film that spontaneously forms on the alloy surface. Ti and its alloys do not provoke allergic reactions and are considered to be highly biocompatible. None of these materials is ferromagnetic. At present, most of the implanted systems are made of pure titanium (CP-Ti) or Ti-6Al-4V alloy. CP-Ti does not yield sufficient hardness for load-bearing applications and is therefore mainly used in dental surgery, for the manufacture of acetabular shells and in coatings for joint replacement. The Ti alloys (Ti-6Al-4V and Ti-6Al-7Nb) exhibit an $\acute{\alpha}$ + β structure and higher compression strength compared to Ti; consequently, they have a broader range of orthopedic applications. However, the wear resistance of Ti and its alloys is relatively low and these materials should not be used where contact wear can occur. Many authors have noted that, although titanium shows a better resistance to corrosion and can be found in living tissue for many years, the release of small amounts of corrosion products (aluminum and vanadium ions from the Ti-6Al-4V implant alloy) is still possible, even through a stable oxide film of TiO_2 and, thus, this layer is not inert [39]. On the other hand, the nanostructure and its further modifications could facilitate and strengthen the implant-bone integration, and thus shorten the healing process.

4. Metal ion release from various metallic medical implants – a systematic analytical review

The discovery of relatively inert metallic and alloy biomaterials has led to their prolific use in biomedicine, such as in orthopedics and dentistry. All metallic prostheses and dental implants degrade to some extent over time, however, resulting in locally and systemically elevated levels of metal ions. Several modes of metal ion release exist, including: passive dissolution, wear (mechanical), corrosion (electrochemical) and combined mechanical and electrochemical processes (e.g. fretting corrosion). The biological risks of metal ions include wear debris, colloidal organometallic complexes, free metal ions and inorganic metal salts or oxide formation [36]. Metal ion release from implants has been reported *in vitro* as well as *in vivo*. *In vitro* tests are quick, simple, enable to assure controlled laboratory conditions, however, do not exactly reflect phenomena occurring in body environment. Metal release can be measured locally, in periprosthetic tissue or more relevantly, in body fluids, i.e. blood, serum or urine, which show the systematic impact of metal release. The comparison of literature data is not often straightforward, since mass transport and metabolism play a role in the accumulation in the tissue *vs.* transfer into the blood or urine [36].

In the sections below, current information on the metal ion release from metallic implant materials and the behavior of released *in vivo* metal ions is presented, according to the empirical data.

4.1. Metal ion levels after total knee replacement (TKA)

Total knee arthroplasty (TKA) is a very successful treatment option for advanced osteoarthritis of the knee. Although patients obviously benefit from joint replacement in terms of mobility and quality of life, implant-specific local and systematic adverse effects, due to corrosion and wear, still constitute a matter of concern [2,4,10].

As they contain large metallic surfaces, TKA implants are particularly subject to corrosion inducing release of metal ions. The literature contains relatively fewer investigations of the ion exposure in total knee implants made of chromium, cobalt, molybdenum and/or titanium alloys [40-48]. Some studies have shown increased concentrations of metal (Co, Cr, Mo and Ni) ions in the serum, urine and synovial fluid [40,41]. Other investigations have shown increased serum, urine and blood metal (Co, Cr, Mo, Ni, Ti, V) ions in loose TKA or patellar components [42-47] but not in well-functioning knee implants. These abnormal metal concentrations proved to be evidence of several kinds of unstable prostheses, which had to be replaced surgically [43].

Rello *et al.* [48] reported a methodology for the determination of trace elements (Mo and Ti) at the μgL^{-1} level in urine samples, in order to monitor metallic implant (metal prostheses) function in humans. The authors considered method offered potential for monitoring patients with metal prostheses in order to detect early evidence of prosthesis failure.

The concentrations of trace metals in the clinical (fluid) samples of patients with ion release from different types of MoM knee joint prostheses are presented in Table 1.

4.2. Metal ion levels after total hip arthroplasty (THA)

A more recent development in the clinical measurement of trace elements relates to the orthopedic area and the increasing use of metal alloys containing Cr, Co, Mo and Ti as the components of MoM hip replacements. The interest in metal ions level following MoM THA and hip resurfacing arthroplasty (HRA) implants started more than three decades ago [49]. The MoM bearing couple is experiencing increased clinical use in both total hip and hip resurfacing applications. These bearings have excellent wear properties, the potential for improved stability and range of motion with larger diameter femoral heads and good midterm results with current generation implants. The most significant disadvantage to these bearings, however, is that metal degradation products are generated, which result in levels of metal in the biological fluids and tissues that exceed levels associated with other bearing surfaces. The level of metal ions in the postoperative period can be used as an indicator of the performance of the bearing surfaces and a marker for ongoing evaluation of implants wear. When evaluating metal ion release arising from MoM THA, one should consider the total load of metal coming not only from bearing wear but also from metallic junctions wear and implant corrosion.

In summary, this section described the total elemental concentration of the metals released by the prosthesis into biological fluids (blood, serum, plasma, erythrocytes and urine) and human organ tissues (kidney and liver) of patients with hip replacements.

4.2.1. Elevated concentrations of trace metals in blood and blood components, urine and human tissues

A number of investigations have been carried out over the years to determine potential metal release into blood and blood components (plasma, serum and erythrocytes) in patients with total hip and resurfacing implant systems. This is certainly not surprising, as blood is one of the biological fluids most frequently used for diagnostic purposes [50,50a].

4.2.1.1. Blood. The measurement of metals (particularly Cr and Co) in the blood can thus be used to provide evidence of bearing couple wear.

In 20 stainless steel Charnley hip arthroplasties, nickel, chromium and manganese levels were measured. Nickel levels in blood, plasma and urine, manganese levels in blood and urine and chromium levels in plasma were significantly higher in the population having received hip prostheses after total joint procedures [51].

Elevated concentrations of metals, including Co, Cr, Ni, Ti and V, have been reported in the blood of patients with both well-functioning and failed total joint replacements [52-86]. Up to 100-fold increase in blood cobalt and chromium levels, from preoperative to postoperative values, has been demonstrated in several series with multiple implant systems. There were also a significant increase in the Ni, Ti and V blood concentrations compared to preoperative values [49,52-58].

Antoniou et al. [59] showed that, at one year after operation, the levels of Co and Cr ions in the whole blood of patients with the 28 or 36-mm-head metal-on-metal prostheses were comparable with the levels observed in patients who underwent hip resurfacing. These results are similar to those described in the study by Daniel et al. [60] and by Vendittoli et al. [61-63], in which the levels of Co and Cr in the whole blood of patients with a Birmingham resurfacing prosthesis were not significantly different from the levels observed in patients with a 28-mm-head Metasul total hip prosthesis [64-67]. Vijaysegaran et al. [64a] reported that in well functioning primary unilateral THA using the Exeter V40 stem with a variety of acetabular components one year post surgery, whole blood chromium levels were within the normal range ($10 - 100$ nmolL^{-1}), with a single mild elevation of serum cobalt (normal <20 nmolL^{-1}). However, the mean whole blood levels of Co and Cr showed a significant increase between the preoperative and one-year postoperative periods [64], and again at six-years post operation [65].

An investigation into the distribution of Co and Cr between whole blood and plasma was reported by Walter et al. [68] in a series of 29 patients with well-functioning unilateral Birmingham hip resurfacing arthroplasty. They concluded that most of the Co and Cr are present in the plasma, and only very minor amounts are associated with red blood cells. Unfortunately, the concentrations reported were just above the normal reference ranges and, because of analytical uncertainties at these low values, firm conclusions on the distribution of these metals are difficult to make.

Smolders et al. [69,70] analyzed the distribution between the serum and whole blood as part of a trial comparing hip resurfacing with metal-on-metal hip arthroplasty. They identified slightly higher Co concentrations in blood than in serum (2.21 nmol/L), and Cr concentrations were lower in blood compared to serum (17.50

nmol/L). Follow-up extended to 2 years and the authors recognized that median ion concentrations in their series were low, with whole blood Co levels of 20.4 nmol/L and whole blood Cr of 21.2 nmol/L. The study provided conversion factors between blood and serum, which may be clinically useful since no consensus exists as to whether blood or serum concentrations should be measured.

Recent evidence shows that metal ion concentrations (Co, Cr and Ni) in the blood are good indicators in differentiating failed from well-functioning prostheses [71-73,73a]. The analysis shows the fundamental differences in the physiological handling of these metals: Co is distributed pretty equally between blood cells and plasma, whereas Cr is mainly in plasma. Additionally, at 3 and 6 months, the analytical methods may allow blood metal levels (Co, Cr)

to be used as realistic biomarkers of the *in vivo* wear rate of metal-on-metal hips [74,75]. The implication being that metal levels can be minimized with optimal orientation of the acetabular component.

Other clinical studies have shown an increasing trend in metal (Co, Cr, Mo, Ni, Ti) ion levels, before and after surgery, in blood over time [76-84,84a,85a, 85b]. Omlor *et al.* [80] monitored blood serum Ti levels in a large cohort of patients who had undergone total hip arthroplasty with a novel modular stem system. From their findings, they concluded that there was no significant increase in serum Ti compared with individuals fitted with non-modular implants. The authors considered that further long-term studies were needed to evaluate the value of serum Ti determination as a diagnostic tool to identify failure of Ti-based implants. Bernstein *et al.* [84,84a] emphasized the need for long-term studies of metal ion release in patients having undergone MoM total hip arthroplasty. The researchers reported that Co and Cr release in blood peaked between 4 and 5 years after operation.

There are concerns related to systemic metal ion concentrations in patients with metal-on-metal bearings and their possible deleterious long-term effects. Pattyn *et al.* [85] conducted a prospective randomized study to compare 3 different metal-on-metal bearings, 2 different resurfacing prostheses and one 28-mm total hip arthroplasty. There was evidence that component positioning, especially the cup abduction angle, was crucial to avoid excessive ion (Co and Cr) release. However, on the other hand, it was concluded that the acetabular inclination angle is not a meaningful determination of metal (Co, Cr) ion levels in Articular Surface Replacement (ASR) system [85a]. Sidaginamale *et al.* [85b] showed that blood/serum metal (Co, Cr) ion concentrations are reliable indicators of abnormal wear processes;

the concentrations of Co in blood are directly correlated to the amount of implant wear.

Lavigne et al. [86] compared the amount of metal ion release (Co, Cr and Ti) from four different types of metal-on-metal prostheses from four different implant manufacturers (Biomet, DePuy, Smith & Nephew and Zimmer). This investigation revealed that metal ion release differs greatly between various total hip arthroplasty implants with large-diameter femoral heads.

The concentrations of Cr, Co, Mo and Ni were measured in erythrocytes of 25 patients with a hip resurfacing implant and compared to the levels in paired serum, erythrocyte and urine samples from 27 controls without an implant. Ion concentrations of Co and Cr were increased in erythrocytes after hip resurfacing arthroplasty [86a].

4.2.1.2. Serum and plasma. Elevated metal ion concentrations in serum and plasma [87-152] have all been reported. The published reports showed that the serum and plasma levels of circulating metal (Cr, Co) ions rose in patients with MoM THA. It has been confirmed that increased ion level is due to the presence of a metal-on-metal implant (coupling prostheses) since ion levels showed a decline following the removal of the MoM implant [113]. In general, studies of serum levels of cobalt and chromium ions have demonstrated variability from patient to patient. The levels tend to be higher in the short term and, with a well-functioning prosthesis, the levels may decrease with time.

There are an increasing number of studies on the level of metal ions (particularly Cr and Co) in body fluids of patients with MoM total hip prostheses. Study groups have assessed ion levels in serum [107-152] that have showed a wide range of Co and Cr, and other metal element (Al, Mn, Mo, Ni, Ti and V) levels, from patient to patient. Metal levels were generally observed to increase in serum with MoM bearings, in those with both the un-operated hips and the metal-on-PE designs.

In a series of papers, Maezawa et al. [108,109,111] and Hasagawa et al. [110] reported that serum concentrations of metal ions (Co and Cr) were significantly increased in patients who had MoM coupling prostheses, even if the implants were well fixed. They also reported that moderate to high serum chromium levels were found in patients who had undergone modern MoM THA after a follow-up period of 3 years. However, not all patients had high serum chromium levels; in fact, some had low chromium levels throughout the observation period. Co levels did not increased for any patients, irrespective of bearing or whether implant was well fixed or loose.

12

Several studies have confirmed an increase of metal ions, Co and Cr and other metals (Al, Mn, Mo, Ni and Ti), in the blood, serum and urine of patients withMoM THA, when compared with healthy individuals, metal-on-PE, ceramic-on-ceramic and ceramic-on-PE THA. The levels tend to be higher in the short term and, with a well-functioning prosthesis, may decrease with time [112-152].

Garbuz *et al.* [150] found substantially higher and clinically concerning levels of serum cobalt and chromium in patients with modular MoM THA with a large-diameter head as compared to hip resurfacing. This implicates the modular MoM contact areas as an important source of metal debris.

In a prospective study, synovial fluid metal levels from stainless steel, cobalt-chromium and titanium alloy cemented total hip implants were measured. Tissue metal levels were quantitated in the cases revised for loosening. Retrieval analysis for implant wear was performed. Synovial fluid analysis showed a fivefold increase in metal levels of loose, compared with well-fixed, stainless steel implants [151].

Nikolaou *et al.* [152] investigated the effect of metal ions on the semen of males of child fathering age with metal-on-metal total hip arthroplasty (MoM THA); cobalt and chromium concentrations were measured in the seminal plasma and in the blood of patients. Results show that the Co level was higher in the seminal plasma of implanted patients (2.89 μgL^{-1}), compared with that of control patients (1.12 μgL^{-1}). This study showed the presence of metal ions in the semen of patients with MoM THA.

4.2.1.3. Urine. Urine metal content is also a promising marker of the ion release in permanent metallic implants.

Elevated urine metal levels in patients with well functioning metallic implants have previously been reported [20, 42, 48, 55, 75, 117, 129, 133] and patients with loose implants were found to generate higher metal levels [53,57,59,66,67,95,103].

4.2.1.4. Human tissues. Contamination was reported, induced by metallic elements (ionic species) on the surface of human tissues, released by joint prostheses [153-160].

Elevated levels of six metals (Ag, Cr, Fe, Mo, Ni and Ta) were found in the tissues directly in contact with the implant ("contact-tissue") [156], and four metals (Co, Cr, Sb and Sc) were additionally found in several organs (heart, kidney, liver and spleen) [157]. The results of Krischak *et al.* [157a] showed that stainless steel plates was more likely to corrode with a markedly higher amount of potentially toxic metallic (Fe, Cr, Mo, Ni) particles release in the soft tissues compared with implants made of commercially pure titanium (cpTi, Ti). Periprosthetic tissue from a group of patients who had failed total hip arthroplasty, in which the prosthesis was made of a titanium alloy, was analyzed [158]. This study demonstrated the continual release of particles of titanium alloy into the tissues, probably from dissolution of extremely high concentrations of the elements (Al, V and Ti) into the surrounding tissues and tissue fluids. Lohmann et al. [158a] hypothesized that the metal (Co, Cr, Ni) content of periprosthetic tissue but not of serum would be predictive of the type of tissue response to metal wear debris. Zeiner et al. [159] developed an analytical method for the determination of seven relevant trace elements (Al, Co, Cr, Mo, Nb, Ni and Ti) in nine kinds of human tissue (brain, heart, kidney, liver, lung, muscle, lymphatic nodes, spleen and body fat) in patients with metallic-metallic THR made of Co-Cr-Mo alloy. The results clearly demonstrated the statistically significant differences in the metal concentrations of persons with and without metallicTHR. The brain and lung are the main target organs for elemental accumulation in persons with hip-endoprostheses and Mo and Nb exhibit the highest tendency to accumulate. Elevated concentration of metal ions (Co, Cr and Mo) were found in the hair of patients with arthroplasties (MoM prostheses) [160].

The concentration of trace metals in the clinical (fluid) samples following ion release from different models of hip joint prostheses is presented in Table 2.

4.3. Metal ion levels from spinal implants

Corrosion affects spinal instrumentation and may cause local and systemic problems; the construction constraints of spinal fixators make them prone to corrosion [161]. Metal particles (wear debris) were generated by the use of titanium spinal instrumentation (pedicle screws) in patients with a pseudarthrosis. The particle remain in the soft tissues; tissue concentrations of titanium were highest in patients with a pseudarthrosis (303.36 $\mu g\ g^{-1}$ of dry tissue); patients with a solid fusion had low concentrations of titanium (0.586 $\mu g\ g^{-1}$ of dry tissue) [162]. In a prospective cohort study [162a], serum samples from 32 paediatric patients, who underwent instrumental spinal arthrodesis, were collected at set times within the first postoperative year; Al, Nb and Ti levels were determined. It has been reported abnormally elevated serum titanium and niobium levels in patients with titanium-based spinal instrumentation up to 12 months.

Elevated levels of nickel and chromium were detected in serum after instrumented spinal arthrodesis with stainless steel implants [163,164]. However, the locking titanium volar distal radius plate did not raise serum titanium levels in patients who had previously received a locked volar distal radius plate [165]. On the other hand, significantly higher serum titanium concentrations were observed in patients with titanium spinal implants (mean, 2.6 µg L^{-1}) when compared with controls (mean, 0.71 µg L^{-1}) [166]. A retrospective study of serum and hair metal concentrations in patients with titanium alloy spinal implants was performed[167]. Approximately one third of patients with titanium alloy spinal implants exhibited abnormal serum or hair titanium concentrations 5 years after surgery. In addition to investigations of patients with artificial hips, other subjects groups with metal implants studied were those with steel bars inserted into the chest to treat *pectus excavatum* (sunken chest) [167a] and metallic spinal lumbar disc replacement [167b]. Metal concentrations were elevated above normal levels by up to 10-fold following surgery and remained increased; blood concentrations of Co and Cr may be more than 100 $µgL^{-1}$ compared with less than 2 $µgL^{-1}$ in healthy subjects.

4.4. Metal ions from dental implants and orthodontic appliances

In dentistry, metallic materials are used as implants in reconstructive oral surgery to replace a single tooth or an array of teeth, or in the fabrication of a dental prosthesis, such as metal plates for complete or partial dentures, crowns and bridges, and are particularly essential for patients requiring hypoallergenic materials.

Depending on the different requirements for the wide range of applications, the dental material market offers a large variety of products; various inert metallic and alloy biomaterials are used in these implant systems [168,168a]. In modern dentistry, the use of dental implants has become the primary treatment regimen for partial and complete edentulism, however, substantial numbers of fixed and removable partial dentures are currently being placed in most patients [169]. No metal or alloy is entirely inert *in vivo*; any metal or alloy implanted in the human body is a potential source of toxicity. Corrosion is one of the possible causes of implant failure after initial success [7,8,170-172]. Due to its mechanical properties, good resistance to corrosion in biological fluids and very low toxicity, titanium has been the most commonly selected material for dental implants and prosthesis [173,174]. It is well known that all dental materials release ions into the oral environment and have the potential to interact with the oral tissues and fluids. However, little attention has been given to metallic ion release from orthodontic implants. During the last 15 years, some remarkable changes in restorative dentistry have occurred. The employment of amalgam and different kinds of alloys has dropped dramatically. The main reasons were connected with the aesthetic aspects and the controversy over amalgam employment and metal toxicity, but also because of environmental pollution from mercury waste. However, there is no convincing evidence pointed out to adverse health effects due to dental amalgam restorations, except in rare instances of an allergic reaction, and can be used as a preferred restorative material where aesthetics is not a concern [175-177].

4.4.1. Dental implant/prosthesis; orthodontic appliances.

Many types of metals and alloys have been used in dentistry. Possibilities for quantitative *in vivo* diagnosis of early dental erosion and monitoring of the treatment efficiencies *in vivo* are very attractive long-term goals. Therefore, clinical monitoring of dental erosion requires systematic and, preferably, quantitative assessment of dental fluids and tissues over a long period of time. Some experiments evaluating the potential for conventional dental implants to release metallic ions into the body have been performed [177a,177b,177c].

It has been reported that saliva leakage between the superstructure (Ni-Cr-Mo alloy) and the implant (made of pure titanium) may trigger a corrosion process (galvanic corrosion) due to differences in electrical potential. This generates the passage of ions such as nickel or chromium from the alloy of a crown or bridge to the peri-implant tissues, with consequent bone reabsorption, which may compromise the mobility of the implant and induce a subsequent fracture [178].

Several *in vivo* studies have demonstrated the corrosion and release of metal ions from orthodontic appliances through emission of electro-galvanic currents, with saliva acting as the medium for continuous erosion over time [179-192].

The release of Cr and Ni from orthodontic appliances into saliva was assessed by Amini *et al.* [179]. Samples collected from 28 subjects with orthodontic appliances (stainless steel brackets, orthodontic bands and archwires) had mean concentrations of 2.6 and 18.5 $ngmL^{-1}$ for Cr and Ni, respectively, while in samples from same-gender siblings without appliances, concentrations were 2.2 and 11.9 $ngmL^{-1}$.

Metal alloy ion release has also been reported [180-192]. Most of these experiments measured metal (Co, Cr, Cu, Mo, Mn Fe, Ni, Ti, V) ion release during the exposure to a biologic medium (blood, serum, urine and mucosa cells) or saliva for periods ranging from 1 day to 1 month. Other studies have also reported the continuing release of metal ions over a 10-month period from a wide variety of dental casting alloys, but the results were not consistent. For example, while some authors have shown an increase in metal ion concentration in the oral fluid of patients with orthodontic appliances [182-187,189,191], Eliades *et al.* [181], Petoumanu *et al.*

16

[189a,190] and Sahoo *et al.* [190a] reported no differences in salivary metal ion concentration between subjects with and without fixed orthodontic appliances. Several authors evaluated Ni and Co [186], Ti, Cr, Mn, Co, Ni, Mo and Fe [186a] and Ni and Cr [186b,186c] levels in oral mucosa cells in the patients with orthodontic appliances (brackets, bands, archwires). In general, it was found that the presence of metals released from orthodontic appliances induced DNA damage and reduced cellubar viability of mucosa cells. Menezes *et al.* [192a] controlled nickel level by another biomarker of exposure, urine. The authors stated that urinary nickel levels increased significantly after 2 months.

Begerow *et al.* [193] investigated to what extent noble metal dental alloys contribute to the total platinum, palladium and gold body burden of the general population. Their *in vivo* investigations confirm the assumption that Pt, Pd and Au are released from noble metal-containing dental alloys by corrosion.

Particles can be released from metallic devices (due to several mechanisms including corrosion, wear and mechanically accelerated electrochemical processes such as fretting corrosion, stress corrosion and corrosion fatigue) and these metallic particles are accumulated and stored in the surrounding tissues [194-196]. Although titanium dental implants are characterized by great biocompatibility, despite the passive activity of the external layer of oxides, both electrochemical and galvanic erosion may take place in the environment of the oral cavity. A remarkable study dealing with the possible role of Ti in the genesis of yellow nail syndrome was discussed by Berglund and Carlmark [194]. Using EDXRF analyses of nail clippings, Ti was regularly found in fingernails of implant patients but not in control subjects. Titanium dissolves from the implants and is deposited into the nails. Meningaud *et al.* [195] conducted a study to investigate whether or not a relationship existed between duration of plating and metal release from Ti miniplates in maxillofacial surgery. Ti was found in the soft tissues in contact with the Ti plates and soluble and non-soluble fractions were distinguished. Their results indicate that almost 100% of Ti is insoluble, most likely corresponding to metallic particles released during the implantation of the plates. Plates, grids and surrounding tissue were investigated to evaluate titanium release and accumulation [196]. The authors concluded that titanium was only present in the interfacial bone, probably due to fretting, and in all fibrous tissue surrounding the devices. High Ti levels were found in blood cells in the connective tissue, however.

The concentration of trace metals in the clinical samples of patients following ion release from different metallic orthodontic appliances are presented in Table 3 and selected operating parameters for trace metal determination in body fluids by analytical atomic spectrometry are presented in Table 4.

4.5. Chemical state and speciation

According to IUPAC [197], speciation analyses are the *"analytical activities of identifying and/or measuring the quantities of one or more individual chemical species in a sample"*, therefore identifying the chemical (oxidation) states of released metals from metal and alloy implants in human tissues and fluids is essential. The cytotoxicity of an element is closely related to the chemical state of the element that has been incorporated in the tissues surrounding the implant. However, when considering the litany of documented toxicities of these elements, it is important to remember that the toxicities generally apply to soluble forms of these elements and may not apply to the chemical species that result from the degradation of metallic implants.

Information about speciation of trace metals released from metallic medical implants is scarce; however, some procedures for speciation have been reported. Interest in speciation based on different sampling conditions appears to be decreasing, with only a handful of publications relevant to this review. There is insufficient interest, however, in identifying the chemical valence state and in measuring the trace metal elements incorporated into a matrix of human body tissues and fluids. For example, chromium from implant alloys may be incorporated into organometallic complexes as Cr^{3+} [Cr(III)] or Cr^{6+} [Cr(VI)]. Since Cr (VI) is far more biologically active than Cr(III), there should be considerable interest in identifying the valence state that predominates in corrosion or wear products, both *in vitro* or *in vivo* [37].

In 1995, Merritt and Brown reported that Cr(VI) is the predominant species of chromium released *in vivo* during corrosion of stainless-steel and Co-Cr alloy implants in the red blood cells (RBCs) of patients undergoing total joint revisions [198]. Thus, corrosion of implants can lead to the release of biologically active hexavalent chromium into the body, which is rapidly reduced to trivalent chromium in the RBCs. They also concluded that the ratio of serum to RBC levels of Cr could be used to measure the proportion of the Cr ion that is in the hexavalent form. Afolaranmi *et al.* [199] confirmed with their study [198] that Cr (VI) is predominantly partitioned into red blood cells, as opposed to plasma. The extent of accumulation in RBCs is influenced by the anticoagulant used to collect the blood, with ethylenediaminetetraacetic acid (EDTA) giving a lower partitioning into red cells compared with sodium citrate and sodium heparin. In contradiction to their study [198,199], Walter *et al.* [68] found that used, normally functioning metal-on-metal bearings showed significantly lower levels of Cr in the RBCs than in the plasma or serum. They concluded that the most likely explanation for most of the Cr being found in the serum is its trivalent form. A probably explanation for the difference between their ratios and those of Merritt and Brown [198] is that most of the ion load from their normally functioning bearings was from wearing, rather than corrosion. A possible explanation for the difference between corrosion and wear products is that the corrosion products are likely to be atomic in size, whereas wear debris, as in their study, is likely to be particulate. Hart *et al.* [199a] determined the chemical form of metal species in tissues surrounding current generation MoM hip arthroplasties. In all MoM hips they found chemical form Cr(III) phosphate (Cr(III)PO$_4$). No evidence of Cr(VI) was seen in the tissues examined.

Very little is known about titanium toxicity regarding the identity of the species [TiO$_2$ nanoparticles or Ti(IV) ions] and their concentrations from orthopedic and dental implants. Nuevo-Ordóñez *el al.* [136] developed and conducted *in vivo* a quantitative Ti speciation method to address the concentration of Ti bound to different human serum biomolecules. It has been observed that Ti is uniquely bound (99.8%) to human serum transferrin. These same authors [130] attempted to explore the possible association of the released metals (Co, Cr, Mn, Mo and Ti) with human serum proteins from the fresh serum of patients carrying MoM (based on Co-Cr alloys) total hip prosthesis and titanium alloy dental implants. Speciation studies have been conducted in samples where the level of Cr, Co and Mn was slightly higher than in control patients. These results revealed the association of Mn to transferrin and of Co to albumin, as well as the fact that Cr could not be detected. It was speculated that Cr might be released from the prosthesis as Cr(VI), which shows no interaction with the studied serum proteins.

Ektessabi *et al.* [200] measured the distribution and chemical states of the trace metal elements (Cr, Fe and Ni) incorporated into a matrix of human tissues. The specimens were from human tissues around a total hip replacement that had been inserted in a patient's body. The particles that originated from mechanical friction between the head and backing of the hip joint system had diffused into the tissue after insertion. The analysis showed that iron and chromium from the stainless steel that were diffused in the tissues had undergone chemical changes; the change in the chemical state of Fe is considered to be due to the dissolution of Fe from the particles of stainless steel.

5. Instrumental techniques and procedures for the measurement of metal ion release from orthopedic and dental implants

For a proper evaluation of implant degradation, damage of implants or enrichment of implant metals in the organism, very low concentrations/amounts of the implant metals have to be determine in human body.

The choice of an appropriate method for analysis depends on several factors: the type and size of the sample, elements to be measured, amount and concentration range, accuracy and precision, speed and cost of analysis, available instrumentation, *etc*. In most clinical studies the number of determined trace metals is limited to macro- and micro-amounts at higher concentrations. Some metals are being investigated in metal-on-metal implant research, although the concentration ranges are considerably lower. The analyzed matrices may include whole blood, serum, synovial fluid, urine and various solid tissues. The measurement of trace metal concentrations for each matrix/metal combination presents its own set of methodological issues. Several of the metals studied are present in body fluids and tissues at concentrations (sub-μgL^{-1}) or amounts (sub-μgkg^{-1}) that are slightly below or slightly higher than their detection limits. Thus, for these elements, the signal-to-noise ratio can be low, making the measured values subject to variability. Thus, the necessity of determining as many metals as possible in different concentrations demands the use of methods and/or a combination of methods permitting multi-element analysis with high accuracy and precision.

The methods most suitable for the rapid and accurate determination of the trace metal content of clinical samples are atomic spectrometric methods, such as atomic absorption spectrometry (AAS), optical emission spectrometry (OES) and elemental mass spectrometry (MS). Methods commonly used in the generation of human body composition data include electrothermal AAS (ET-AAS), also known as graphite furnace AAS (GF-AAS), inductively coupled plasma (ICP-OES) and ICP-MS. Although ICP-OES can be, and sporadically is, successfully used for the determination of trace metals in human body fluids, it is not discussed in this paper, because of its relatively limited applicability and the fact that these metals are also easily measured with more robust techniques, such as ET/GF-AAS and ICP-MS.

In this section, the application of analytical instrumental techniques for measuring the total elemental concentration was described for the metals released by the metallic medical implants into biological fluids (blood, serum, plasma, erythrocytes, urine and saliva) and biological organ tissues (kidney and liver) of the patients with MoM prostheses and metal dental implants.

It is difficult to compare data related to the performance of metal implants, due to a large variety of laboratory techniques, media used for measurements, implant metal materials, patient demographic data, follow-up periods, type of study, *etc*. In general, two major analytical techniques are used, ET-AAS (GF-AAS) AAS and ICP-MS.

5.1. Atomic absorption spectrometry (AAS)

Atomic absorption spectrometry (AAS) is a "classic" analytical technique for the determination of numerous trace metals in clinical and biological samples; it is well established and well documented, widely available, affordable, robust and reliable, although its gradual decline and replacement with ICP-MS is inevitable. The main limitation of the method is that it is typically a single-element technique, although continuum-source and sequential AAS systems have been developed.

Digestion of solid or liquid samples is typically needed for most ET/GF-AAS applications, because liquids are easier to homogenize, dilute, handle with autosamplers and introduce into atomizers than solids.

Cobalt, Cr and Ni concentrations were determined by ET-AAS in serum and urine specimens collected from a group of 28 patients at intervals of 1 day to 2.5 years after TKA with porous-coated prostheses fabricated of a Co-Cr alloy [40].

Samples of synovial fluid and blood were analyzed for Co, Cr and Mo concentrations, using GF-AAS. NIST SRM 909 Human Serum containing 91.3 μgL^{-1} chromium was used as the Cr Standard [41].

Leopold *et al.* [42] developed a method to quantitatively determine Ti in serum by GF-AAS. Serum metal ion (Ti) levels were used in the diagnosis of a failed TKA.

Samples of blood were analyzed for Cr, Co, Mo and Ni [43] and Co, Cr and Mo content [45] using GF-AAS.

Titanium, Co and Cr element concentrations were determined in whole blood and urine specimens from patients with cementless TKA by ET-AAS [47].

Rello *et al.* [48] developed a method for the direct and simultaneous determination of Mo and Ti in the urine of patients with metallic prosthesis, after its deposition onto clinical filter paper, giving rise to a dried urine spot (DUS). The measurements of Mo and Ti in DUS were carried out using solid sampling high-resolution continuum source GF-AAS (SS HR-CS GF-AAS).

Generally, ET-AAS measures metal ion levels in body fluids, such as in blood [53-55, 57-60,68,70,73,74,82,93,120,122,130], serum [68,93,100,108,115-120,122,125, 137,139,141], urine [53,55,57,60,70,73,93,100,114,126,127,174], plasma [68,73], and saliva [179, 181,183-185,191], as well as in tissues [182,186,187,195].

Selected operating parameters for trace metal determination in clinical (body fluid) samples of patients with detection by ET-AAS are presented in Table 5.

5.2. Inductively coupled plasma - mass spectrometry (ICP-MS)

In recent decades inductively coupled plasma-mass spectrometry (ICP-MS) has emerged as the most promising technique for multielement trace/ultratrace metal analysis of clinical and biological samples. The ICP-MS provides high-performance analysis: rapid (1 – 2 min) compared to tens of minutes with AAS with high sensitivity. Despite several advantages, ICP-MS is usually applied for a rather limited number of metals, considering its capabilities, probably due to the research aims in the respective studies.

The main components (Ti, V, Cr, Co, Ni and Mo) of metallic alloys currently used in knee articular prostheses have been simultaneously determined in human whole blood and urine of implanted people by a high resolution ICP-MS (HR-ICP-MS) [46].

Metal ion concentrations have been measured in patients with MoM hip arthroplasties. Elevated metal ion concentrations in whole blood [46,56-61,63,64,66,68,69,72-75,78-86,110,122,126,130,131,216], serum [68,69,104,106,110,114,119,120,122,125,127,128,131,136,139,147,150,160,217], plasma [73], urine [46,57,60,63,73,99,114,126,130,133,138,139,143,160,163,166,167,224], saliva [70,181,186,187,190,192,193], mucosa cell [186,182a,186a,b,c,189a], tissues [186] and hair [167] have all been reported.

Selected operating parameters for trace metal determination in clinical (body fluid) samples of patients with detection by ICP-MS and ICP-OES are presented in Table 6.

6. Sample collection and preparation of biological fluids and tissues

6.1. Contamination control

Regardless of whether AAS, ICP-OES or ICP-MS techniques are used, contamination is still a major technical challenge in trace, and especially ultratrace,

22

metal analysis, particularly when levels in the μgL^{-1} or sub-μgL^{-1} range are measured. In order to generate high quality analytical data at ultra-trace levels, all potential sources of contamination must be identified and, as far as humanly possible, eliminated [201,202].

Metal contamination can produce erroneously high results, because of the addition of impurities to the sample. Such impurities are usually introduced by the process of containment, contamination during aliquoting, absorption of air-borne contaminants or by reagents used in processing the specimen before and during analysis. The reagents used throughout the sample preparation and analysis provide potential sources of contamination and must be carefully evaluated. It is imperative to maintain full control over the high-purity reagents and water used for sample preparation, in order to minimize potential adulteration of the analysis. For trace and ultra-trace metal applications, several companies offer ultrahigh-purity reagents, as well as high-purity acids and bases for biological and clinical applications. Adsorption of the trace metals onto container walls could be a type of contamination through loss. Contamination control begins with specimen collection. Selecting an appropriate specimen collection device is critical to success. Simply using standard biological fluid collection tubes will render the results useless. Several manufacturers market fluid collection tubes specifically for trace element analysis and some certify tubes for specific element(s). Contamination may originate from the tube stoppers, e.g. Zn, from the anticoagulant, or even from the container material, such as Al in glass. Thus, it is good practice to check that collection devices are free of significant contamination specific to the element(s) to be measured. A plastic needle, rather than a metal needle, for the venipuncture could avoid contamination with elements from the needle alloy [203]. If the needles are made of stainless steel, there is a potential hazard of contamination with Fe, Cr and Ni. Handling of the samples should therefore be kept to a minimum and metal-free gloves should be worn.

During the sample processing stage, contamination can generally be controlled with good laboratory practices. In an ideal situation, clinical samples would be processed in high-energy particulate air (HEPA) filtered clean rooms, designed to have minimal contamination. Clean hoods and laminar airflow units also prove useful.

6.2. Sample collection and storage

A great number of serious errors may be committed during the range of procedures from sample collection and storage to the ultimate detection of the metal

[202,204,205]. Specimens of biological fluids for trace and ultra-trace metal analysis are usually obtained by puncture. The low concentrations in which trace elements are present in body fluids demand the utmost care in collecting a specimen, particularly if the interpretative and diagnostic value of the analytical data is to remain inviolate. Changes in the trace element concentrations of biological fluids and/or tissues during storage are usually due to contamination by desorption from the container or loss by adsorption to the container. These can be minimized by using sample bottles with large volume-to-surface ratios made of Teflon (polytetrafluorethylene (PTFE), perfluoralkoxy (PFA), tetrafluormetoxil (TFM) or less expensive PE and polypropylene containers that have been thoroughly acid washed prior to use. The use of anticoagulants is very problematic, as most anticoagulants are either polyanions (e.g. heparin) or metal chelators (e.g. EDTA, citrate) and therefore have a high affinity for metal ions. In lieu of chemical preservation, the sample should be stored either in a refrigerator at <5 C° or frozen at -20 C°, or less (plastic tubes only), until needed. The length of time that the sample can be stored without changes occurring is element-dependent, although the analytical uncertainty for most elements is too high to discern any difference during reasonable periods of storage. Lyophilization prior to freezing may reduce or eliminate such changes.

A particular challenge arises in the biological sciences when the investigated systems are highly complex and dynamic. The requirement for minimal interference with the biological system is best met by microanalytical techniques, based on microliter- and nanoliter-sized samples, or even *in vivo* sample preparation [206]. Microsampling techniques are poised to revolutionize sample collection and preparation for *in vivo* and *in situ* studies. Their small scale enables convenient *in vivo* sampling and a significant reduction in overall analysis time and reagent use.

6.3. Sample preparation: dilution, dissolution and digestion

In atomic spectroscopy (i.e., AAS, ICP-OES and ICP-MS), the choice of the sample preparation method is one of the most critical steps in the determination of metals concentrations. The selection of the method involves many important points: *(i)* the nature of the clinical sample; *(ii)* the analytical technique employed for the analyte (metal) determination; *(iii)* the number of samples; *(iv)* the analyte and its range of concentration; *(v)* the trueness and precision required; and *(vi)* the sample pre-treatment time. Due to the high viscosity of some body fluids, they cannot be directly introduced into the atomization/excitation/ionization source hence requiring from a previous sample treatment. The main preparation methods are: *(i)* dilution; *(ii)* dissolution; and *(iii)* decomposion/digestion.

Dilution is important for direct measurement of aqueous samples such as serum, plasma, urine, *etc.,* especially for serum and plasma, as they have a high salt and protein content and high viscosity. Dilution with an appropriate solvent is a straightforward, fast and accurate procedure; water, acids, Triton X-100 and TMAH are mainly used. Furthermore, it is very easy to automate.

Dissolution and digestion of samples are other important steps of the analytical techniques, which are sources of contamination and losses. The most important sample digestion techniques are dry ashing and wet digestion. Dry ashing involves destroying the organic matrix by combustion and dissolving the trace metals from the ash. Samples prepared by dry ashing are vulnerable to the loss of trace metals by volatilization. For this reason, ignition temperatures are maintained below 450 C° or ashing aids (e.g. magnesium nitrate) and modification of the sample matrix are often used to eliminate potential losses. However, these can also lead to contamination of the sample and poorer detection limits.

Sample wet digestion (decomposition) consists in the conversion of the organic matrix in an aqueous one and is used for digestion in open vessels, in closed pressurized vessels and with automated flow systems at an elevated temperature. The organic matrix is destroyed by oxidation with the acid, which dissolves the inorganic residue. Nitric acid (HNO_3), alone or in combination with perchloric ($HClO_4$) acid or hydrogen peroxide (H_2O_2), has frequently been used for this purpose. The most convenient acid for spectrometric applications is the HNO_3 because no significant analytical problems have been observed with its use at concentrations up to ~10%. H_2O_2, often added to mineralization mixtures, is also rarely responsible for analytical problems, whereas the use of HCl_4 may be problematic with GF-AAS. HNO_3 is usually the choice for ICP-MS analysis, as it causes minimal interferences. Wet digestion in high-pressure decomposition (closed) vessels has the advantages of reduced losses by volatilization and reduced contamination by reagents.

6.4. Quality control considerations

The question of quality control in analytical determinations is of vital importance, especially when a large number of metals is to be determined. Once the method has been developed, it must be validated, ensuring that the method produces data in agreement with the "true value" of the metal in the samples. This can be accomplished by running a reference material, a major tool for developing and validating analytical methods, to see if the method is producing accurate results. The clinical Certified Reference Materials (CRMs), Standard Reference Materials (SRMs) or Reference Materials (RMs) with certified or information values for numerous metals are available, which provides a list of samples: organs, tissues, human body fluids (whole blood, serum, urine, dialysis fluid) [207-209]. Reference materials are generally provided by government-run laboratories with expertise in the measurement of trace elements, and who provide a certificate stating the origin of the materials, the procedures by which values are certified and a statement of the analytical uncertainty. The usefulness of an RM is highly dependent on its similarity to a specific test, not only should the homogeneity and stability of the material be considered, but also the level of metal concentration, the chemical form of the metal of interest (in the case of speciation), the matrix match and potential interferences, the measurement uncertainty and the value assignment procedure. Although use of CRMs, SRMs or RMs is recommended to verify accuracy, they are usually treated (e.g. lyophilized) for long-term conservation, and therefore may behave differently from real patient samples. Reference materials are always "idealized" matrices and as such relatively easy to handle. Thus, the analyst should be cautious in interpreting CRM/SRM/RM results and reporting of clinical studies.

Control materials from commercial sources, for example Seronorm (Trace elements whole blood, serum, urine) Nycomed Pharma AS (Norway), may be less well characterized, and "expected" values may not always be reliable, or they may be accompanied by very large uncertainties.

Table 7 provides an overview of the trace elements that can be found either as certified or information values in different CRMs/SRMs or RMs or as commercial control materials, and links to the respective websites.

6.5. Units for metal concentrations in clinical samples

A variety of units are used to report metal ion concentrations in clinical samples [210]. While many researchers and much of the clinical laboratory community around the world use substance ($nmolL^{-1}$ or $\mu molL^{-1}$) concentrations, there is no such consensus within the USA, where a myriad of mass (μgL^{-1}, mgL^{-1}, $\mu g/dL^{-1}$, $ngmL^{-1}$, ppm, ppb, *etc.*) concentrations are more frequently used. Mass concentration units are frequently preferred within the analytical laboratory.

7. Conclusions and future prospects

The answer to whether or not the corrosion of metal surgical constructions and dental implants are clinically important phenomena lies in a consideration of the various activities of metal particles/ions release *in vivo,* which can permeate every tissue in the body of a metallic implant patient. Metal-based implants undergo corrosion and wear, generating metallic debris that can exist in several forms, including organometallic complexes, free metallic ions, inorganic metal oxides and occasionally nanoparticles. Metallic implants (orthopedic and dental) or wear debris generated from implants, may release chemically active metal ions into the surroundings tissues. Although these ions may stay bound to local tissues, metal ions also may bind to protein moieties that then are transported in the blood stream and/or lymphatics to remote organs. In an effort to reduce the amount of ion release from wear of articulation, the science of tribology has shown that MoM components made of high carbon content Cr-Co with small clearance, a good surface finish, and optimal implant sphericity can minimize ions release from bearing wear and thus osteolysis. Evaluation of body fluids and tissues around metallic devices is important, since the presence of ions/particles and their potential local biological effects might affect implant success. Moreover, the understanding of implant behavior requires the analysis of failed surgical and dental implants with nearby body fluids and tissues. The main goal is to obtain correct results in the determination of trace metals and to provide and substantiate the reliability and validity of these results. The determination of trace and ultra-trace metals in the body is a very complex problem. This status depends on many factors: the intake of trace metals, their absorption and excretion, the bioavailability of the trace metals and their homeostatic control.

The original purpose of measurement is to compare results. Furthermore, the most important aim is to reach comparability of the results beyond the limits of one laboratory (inter-laboratory reproducibility), so that generally valid conclusions can be made with regard to physiological and pathological processes. One of the general limitations in the literature regarding metal ion concentrations is the wide variability in how measurements are gathered, analyzed, and reported, thus making comparisons between studies difficult. However, the measurement of very low levels may be more critical, and further investigations may be required in view of the differences observed between the laboratories. Variability among different laboratories is attributed to multiple factors; potential for contamination is one important factor to be prevented through choosing the right sampling technique.

It is evident that considerable research has been undertaken during the last several decades in the area of sampling pretreated body fluids and tissues, combined with atomic and mass spectrometric detection. Fragmentary reviews published during the last ten years evidence the continued interest in the influence of metallic medical implants in the human body, regarding hip arthroplasties [3,34,50,211-215] and orthodontic appliances [177a,177b,177c,214a]. However, in terms of dental implants, since continuous monitoring and assessment of metal levels in saliva is not currently available, several issues must be resolved before any definitive conclusions can be drawn on the potential for metal release from orthodontic appliances *in vivo*. Continuous monitoring and assessment of metal ion levels in saliva could help to understand the corrosive potential of orthodontic appliances *in vivo*. Thus, *in vivo* investigations are urgently needed to study the behavior and biocompatibility of different commercially available dental alloys under real-life conditions. From the interpretation of the literature data, it is obvious that the results reported for *in vivo* bio-accessible fractions of different released trace metals vary greatly from study to study. Exposure should be carefully monitored to clarify the biological effects of ion dissemination and to identify risks concerning the long-term toxicity of metals. It would be fair to say that, in general, most of the fundamental parameters and requirements for analytical procedures have already been established and virtually all of the work examined in this review principally concerns analytical applications (summarized in Tables 1 - 6). However, the compilation of obtained results from different clinical laboratories shows that the accuracy of trace, and especially ultra-trace, metal determinations is often inadequate and gross systematic errors occur quite frequently, because very low values can be of clinical importance for the evaluation of implant performance. Possible explanations include potential for contamination during sampling, contamination during storage or contamination during sample handling. Environmental contaminants can come from dust in the air, air pollution from factories that release metal particles, contact with metal surfaces and cigarette smoke. Therefore, special care should be given to minimizing airborne contamination, strict decontamination methods, proper hand hygiene selection of uncontaminated sampling equipment and monitoring the laboratory for contamination.

For the development of new implant materials and investigation of risk potential for already applied materials systems we do need a better understanding of the *in vitro*, *in vivo* and *ex vivo* (investigation of *in vivo* aged samples) corrosion process of these implant alloys. Investigation of these processes makes the application of highly sensitive analytical methods with high lateral resolution necessary. With these methods detailed investigations of the only few μm thick corrosion zones around the implant and of the changes in elemental concentrations in the surrounding tissues as well as other organs will be possible. In the focus of such measurements are the determination of the composition of the solid degradation products, the ion or particle release with time and the interaction of these materials with biomolecules in the tissue.

In addition to the problems associated with sampling and sample pretreatment, the determination of released trace metal fractions using parts per billion (μgL^{-1}, μgg^{-1}), or an even smaller level (sub-$\mu gL^{-1}/\mu gg^{-1}$), as a measure is also analytically challenging, since accurate analysis is hampered by small sample amounts, technical complexities of working with nanometer-sized particles and by the presence of a highly complex clinical sample matrix, mainly derived from the type of chemical agents used. These factors can cause matrix interferences and insufficient sensitivity, if standardized routine procedures (ET-AAS, ICP-MS and, sporadically, ICP-OES) were the main detection tests used for analysis. For example, acid digestion of clinical samples is necessary to determine total metal ion concentrations accurately so that meaningful comparisons can be made between clinical studies adhering to a standardized laboratory protocol, thereby providing a more accurate reflection of actual concentrations. Thus, the use of inadequate analytical procedures or a change in the laboratories measurement performance might also contribute to the observed variations for bio-accessible trace metal fractions in body fluids and tissues. To establish what might be an acceptable level of trace metals guidelines require to be established to standardize analytical methods (measurement of ion concentrations), collection techniques and standard values and bioavailability in this complex subject. To overcome the limitations of routine methods, improved procedures were developed, applying special strategies for sample introduction, separation of interferences or signal quantification. In orthopedic surgery, metal ion determination of body fluids can be performed in blood, serum, knee and hip fluids and urine. The results of the study of Rodríguez de la Flor et al. [160] show that hair is a good biological marker for the monitoring and study of the toxicokinetic behavior of metals released from MoM hip prostheses, and as such it may have immediate clinical applications for mid- and long-term clinical follow-up. At present, based on literature data, measurement of trace metals in whole blood is most practicable and more accurate and favored rather than the level in plasmas, serums or RBCs and the measurement of serum and whole blood metal concentrations should not be used interchangeably or interconvertibly. This should be useful for surgeons because blood metal ion levels can be used as an *in vivo* biomarker of wear rate, eliminating the

29

need for more invasive joint fluid analysis [216]. However, there is no consensus on which matrix (whole blood or serum) is superior, although, both matrices are used in routine clinical practice. Current literature focuses on blood concentrations and consideration should be given to also measuring trace metals in urine to monitor the effect of increased outputs on the kidney. 24 h urine metal concentrations are more reliable, but a 24 h urine collection is cumbersome and often incomplete. Hip fluid concentrations may also be informative when whole blood or serum metal concentrations are not conclusive. Cobalt and chromium should also be monitored as a reference metals. The measurement of the Co and Cr ions both in whole blood and in serum can be considered a standard test for screening and monitoring patients with MoM since it is strictly correlated with the release of wear particles and corrosion products at the local level. In general, a cut-off level of 7 μgL^{-1} for either cobalt or chromium level as a predictor of failure of MoM hip, no other orthopedic implants, with 90% specificity but only 50% sensitivity was used. In 2010 the UK Medicines and Healthcare Products Regulatory Agency (MHRA) adopted this level as a marker for their safety alert. Although, currently a threshold for cobalt and chromium between 4 μgL^{-1} and 7 μgL^{-1} is under debate [216a]. Very high levels (>20 μgL^{-1}) or steady increase over time should be a warning sign.

The importance of X-ray fluorescence (XRF) in clinical chemistry cannot be overlooked and is a promising area for development. The technique has been applied to the analysis of body fluids and tissues, but it is for *in vivo* measurements that XRF is particularly valuable in clinical chemistry. Specimen collection is highly invasive and difficulties with calibration and complexities of instrumentation have been impediments to the rapid development of *in vivo* XRF.

Recent advances show that plasma MS and ET-AAS can meet the current needs for low-level trace metal analysis in various clinical and biological matrices [216b]. The main analytical problem in determining these metals in body fluids and tissues is they are present at extremely low (sub-μgL^{-1}; sub-μgkg^{-1}) concentrations/amounts, in a very complicated matrix (blood, serum, plasma, urine, tissue). With appropriate contamination precautions, ET-AAS has been successfully used for such analyses, but one of its disadvantages is that the analytical procedures are relatively time-consuming. However, the method obviously benefits from the performance of HR-CS GFAAS instrumentation, which permits the simultaneous measurement of the target analytes in contrast with "classic" line source AAS and provide an interesting opportunity to see whether this capability will find an application in routine clinical laboratories. Thus, for the technology is very new and so experience with clinical applications is limited. In addition, some of the concentrations reported for metals were below or very close to the detection limit of the ET-AAS technique and, in most cases, below the limit of quantification. Therefore, it appears that ET-AAS is not a consistently reliable analytical technique to measure metals in blood, serum, plasma or urine at μgL^{-1} (parts per trillion) levels. In light of this, some of the reported results using ET-AAS to measure levels at parts per trillion should be revised. Even AAS is still the dominant analytical technique used for trace element analysis in clinical laboratories, more and more clinical laboratories are transitioning away from ET-AAS techniques toward those based on ICP-MS. However, mature techniques such as flame AAS and ET-AAS will continue to be used in situations where they provide adequate results and where the techniques are "fit-for-purpose". In spite of that, ICP-MS, with the possibility for simultaneous multi-element determination, is emerging as a powerful alternative and as the cost of ICP-MS instrumentation becomes more affordable, more routine clinical laboratories are adopting the technology. Although, usually procedures of evaluation are carried out in external, highly specialized laboratories. Newer magnetic sector ICP-MS (HR-ICP-MS) instruments are potentially very powerful tools for the analysis of samples with complex matrices, is more accurate and more reliable than conventional MS and GF-AAS. HR-ICP-MS instruments routinely have detection limits one order of magnitude lower than quadrupole ICP-MS instruments, but current instrumentation may be too costly for routine clinical laboratories. From a clinical point of view, HR-ICP-MS has acceptable clinical repeatability, is extremely robust and can rapidly deliver highly accurate measurements for multi-elements at low concentrations in difficult matrices such as whole blood, serum, synovial fluid or urine. Development of atomic mass spectrometers and searching for a "better ICP" that are more efficient in the detection of atoms and which use more of the sample will be required in the future, such as double-focusing sector field (SF-ICP-MS) [217], dynamic reaction cell (DRC-ICP-MS) [217a], triple-quadrupole ICP-MS/MS instrument [217b] and/or ICP-MS instrument in MS/MS mode [218]. In addition, because the ICP requires lots of power, lots of argon, and generally substantial amounts of sample material, there should be studies on small sources that operate in the open air or with air as the support gas, at modest power levels, that require only small amounts of sample, but which rival ICP performance could be microwave induced plasma (MIP).

Over the past years, the methods based on these two techniques (ET-AAS and ICP-MS) have evolved enormously. Thus, the present analytical review clearly shows that initially procedures such as clinical sample dilution or decomposition should switch to a more direct sample analysis methods. Nowadays there are liquid sample introduction devices available (*i.e.,* low sample consumption) appropriate for the direct introduction of liquid solutions into the plasma with a minimum dilution factor. In fact, the dilution factor has decreased from 1:100 or 1:50, recommended when conventional sample introduction systems are used, to 1:1 for micro-sample introduction systems. With these developments, it is possible to envisage a significant improvement in the detection limits not only due to a decrease in the sample dilution factor, but also caused by an enhancement in the efficiency of the system performance. Another interesting point should be related to alternative systems for solid (*i.e.,* human soft tissue) sample introduction in ICP-OES and ICP-MS. Electrothermal vaporization (ETV) and laser ablation (LA) methods for clinical analysis should be promising; they circumvent some of the problems related with high solvent plasma load or sample dilution [218a]. The combination of spectrometers and microscopes should benefit from the use of microscope objectives and should allow mapping of larger areas and, consequently, the distinction of local variations. As one would expect there is also considerable interest in real-time detection *in vivo*, with specific reference to nanoscale analysis. Future directions for atomic spectrometry in clinical laboratory medicine should include the increased use of isotope ratio measurements for source identification purposes. Isotopically enriched elements can be used as tracers to study mechanisms of absorption and distribution in the body. Although speciation measurements are currently limited to academic and government research laboratories, they will likely be offered routinely by commercial clinical laboratories in the next few years, as method validation improves and robust reference interval for species is established.

The importance of ICP-MS to the clinical chemistry of trace metals is still being realized. Successful coupling to a separation system, such as a chromatography column, high performance liquid chromatographic (HPLC) systems or capillary electrophoresis (CE) has made elemental speciation procedures much more effective. Also offers the advantage for on-line speciation with very low detection limits. The power of elemental speciation using plasma-related techniques is based on hyphenated methods that allow for the performance of time-resolved analysis. Current technologic tools (GF AAS and ICP-MS) can measure only the total element concentrations; as opposed to an element speciation analysis. Despite the analytical capabilities of ICP techniques, however, the analysis of total metals in the human body is currently more popular, by a large margin, than elemental speciation. In the body, element (metal ion) is present in various forms all of which have different biological functions. A meaningful risk assessment, then, should require speciation analysis. An example of this is to use typical speciation strategies (HPLC-ICP-MS) in order to discover ligands in biological fluids. Only a few publications deal with speciation of the degradation products of metallic joint replacement prostheses, making it difficult to draw any significant conclusions on this matter, although encouraging preliminary results have been reported [130,131,136,199]. The degree of harmful toxicity will not only depend on the concentration, but also on the actual chemical form and exposure time. Such studies represent an enormous challenge because of the technical complexities of working with nanometer-sized particles and ion concentrations in the parts per billion range. Considerable work is required to discern the chemical form of the metal, the nature of its ligands and, ultimately, its potential toxicity. Ultimately, specific toxicologic investigation of relevant species can be used in animal models and cell cultures to delineate the biologic effects of these degradation products. For speciation study a molecular mass spectrometric technique, such as electrospray ionization time-of-flight mass spectrometry (ESI-TOF-MS) or matrix-assisted laser desorption/ionization mass spectrometry MALDI-MS, can be used to complement separation methods coupled to ICP-MS.

Once a method has been developed it must be validated, ensuring that it produces data in agreement with the true value of the metal in the samples. This is accomplished by calibrating with the method of standard additions (to check for matrix interference effects), analyzing the given sample with two or more analytical techniques based on various physicochemical principles or by participating in an interlaboratory comparison. One of the best validation methods is the analysis of reference materials (CRM/SRM/RM) of the matrix type and the concentration level of the metals close to those in the tested clinical samples, in order to see whether or not the method is providing accurate results. It is also very important to investigate and interpret the clinically relevant fractions of trace metals. Different forms of metals may have very different biological properties, and measurement of the total concentration may not provide accurate information concerning the clinical status. Therefore, there is still an urgent need for certified clinical RMs that contain endogenously bound trace metals certified at clinically-relevant concentrations, in order to validate speciation methods. There are just a few CRMs available that have certified concentrations of specific elemental species, and thus robust method validation are difficult to accomplish. Currently no whole blood, serum or urine reference materials with certified concentrations of trace metals (Co, Cr, Mo, Ni *etc.*) of unexposed individuals are commercially available. The materials that exist are

33

either (1) not certified, (2) provide values at (non-certified) elevated concentrations suitable for occupational health studies only, or (3) provide only a certified value for a single element of interest, i.e. Co in NIST 2670a Toxic metals in urine. However, the availability of clinical CRMs specifically for speciation methods is expected to improve.

Besides their use in orthopedic and dental implants, some common uses of nanoparticles are as a white pigment in food coloring, toothpaste, even in "natural supplements" and mineral makeup and as an ultraviolet block (sunscreens) in cosmetics in the form of titanium dioxide (TiO_2) and zinc oxide (ZnO), which are pulverized down to nanoparticle size. These particles are small enough to pass through the skin wall and even enter into cells of the body (mainly in the bloodstream), tissues and organs after uptake by the gastrointestinal tract or via human inhalation exposure [219]. In addition, concerns regarding elevated Ti ions in the blood arise from the ability of Ti dioxide, an agent commonly used in air and water purification, to induce pulmonary toxicity when inhaled. In 2006, the International Agency for Research on Cancer (IARC) classified pigment-grade TiO_2 as a group 2B carcinogen (possibly carcinogenic to humans) although in 1999, Ti metal implants (alloys) were considered group 3 carcinogens (not classifiable as to their carcinogenicity to humans). Very little is known, however, about Ti toxicity regarding the identity of the released species [TiO_2 nanoparticles or Ti(IV) ions] and their concentration. Therefore, the strict control of the Ti and Zn levels in the fluids and tissues of the exposed population is of general interest, in order to evaluate possible pathological conditions associated with concentration increments. For this purpose, the first step should be the establishment of Ti and Zn concentrations in non-exposed individuals, which would permit the marking of a threshold between natural Ti and Zn concentration variations and levels of exposure. In addition, the average dietary intake of these metals (Ti and Zn) and others (Al, Co, Cr, Fe, Mn and Ni) and their intake through commercial multivitamin/mineral dietary supplements, cosmetics, pharmaceuticals and food and drink should be established [220-223]. It would be advantages that the subjects (patients) would fill a questionnaire which would include questions about other potential sources of metals potentially released from metallic implants in their organism. Finally, it should be stressed that the correct interpretation of the systemic metal ion concentrations implies the exclusion of other sources of metal ions, i.e. Co and Cr are present as stabilizers and coloring agents in many rubber products, including the colored stoppers on some evacuated blood-collection tubes and in most disposable syringes.

In conclusion, there is no indication for therapy without reliable diagnostics. It is the task of the analytic disciplines to make diagnostics so safe and practical that therapy will be given a healthy foundation. This must be our common goal with regard to trace elements (metals), as in everything. Therefore, metal ions monitoring (a single measurement) or trend characteristic in metal ion levels [224,224a], especially in at high-risk patients, using appropriate technique of sample collection and analysis are vital step in monitoring metallic implant function and should have sequential metal ion level testing as part of the routine follow-up. In the field of medical diagnosis, the concept of being able to probe a tissue *in situ* rather than waiting several days for the result of a biopsy could revolutionize dental and orthopedic surgery.

This review paper seeks to provide a guideline to various aspects of metal ion analysis in patients with metallic medical implants. These include contamination control, specimen collection and storage, sample pre-treatment, analysis (determination metal ions after the implantation of any kind of artificial implant), standardization of the methodology and data presentation (standard values). Metal ion concentrations cannot be used as the sole parameter and an isolated screening tool, however, and must be interpreted as an adjunct to a thorough clinical evaluation [225], radiological and cross-sectional imaging, metal artifact reduction sequence (MARS), magnetic resonance imaging (MRI), ultrasound and computed tomography (CT) findings.

Reassuming, to achieve reliable and comparable results, *in vivo* and *ex vivo* tests should be carried out under strictly controlled conditions so the results would be reproducible and meaningful. There is a need to elaborate certain rules and standards for *in vivo/ex vivo* studies in this area of interest.

8. References

[1] Virtanen S, Milošev I, Gomez-Barrena E, Trebše R, Salo J, Konttinen YT. Special modes of corrosion under physiological and simulated physiological conditions. Acta Biomaterialia 2008;4:468-76.

[2] Kamachi Mudali U, Sridhar TM, Ray B. Corrosion of bio implants. Sādhanā 2003;28:601-37.

[3] Hanawa T. Metal ion release from metal implants. Mater Sci Eng 2004;C24:745-52.

[4] Cadosch D, Chan E, Gautschi OP, Filgueira L. *Metal is not inert*: Role of metal ions released by biocorrosion in aseptic loosening – Current concepts. J Biomed Mater Res Part A 2009;91:1252-62.

[5] Singh R, Dahotre NB. Corrosion degradation and prevention by surface modification of biometallic materials. J Mater Sci: Mater Med 2007;18:725-51.

[5a] Steinemann SG. Metal implants and surface reactions. Injury (Suppl) 1996;S:C16-22.

[6] Kirkland NT, Birbilis N, Staiger MP. Assessing the corrosion of biodegradable magnesium implants: A critical review of current methodologies and their limitations. Acta Biomaterialia 2012;8:925-36.

[7] Olmedo DG, Tasat DR, Duffo G, Guglielmotti MB, Cabrini RL. The issue of corrosion in dental implants: A review. Acta Odontol Latinoam 2009;22:3-9.

[8] Bhola R, Bhola SM, Mishra B, Olson DL. Corrosion in titanium dental implants/prostheses – A review. Trends Biomater Artif Organs 2011;25:34-6.

[9] Antunes RA, Lopes de Oliveira MC. Corrosion fatigue of biomedical metallic alloys: Mechanisms and mitigation. Acta Biomaterialia 2012;8:937-62.

[10] Bischoff UW, Freeman MAR, Smith D, Tuke MA, Gregson PJ. Wear induced by motion between bone and titanium or cobalt-chrome alloys. J Bone Joint Surg 1994;76-B:713-6.

[11] Jacobs JJ, Gilbert JL, Urban RM. Current concepts review. Corrosion of metal orthopaedic implants. J Bone Joint Surg 1998;80-A:268-82.

[12] Urban RM, Jacobs JJ, Gilbert JL, Galante JO. Migration of corrosion products from modular hip prostheses. J Bone Joint Surg 1994;76-A:1345-59.

[13] Daley B, Doherty AT, Fairman B, Case CP. Wear debris from hip or knee replacements causes chromosomal damage in human cells in tissue culture. J Bone Joint Surg 2004;86-B:598-606.

[14] Milošev I, Remškar M. *In vivo* production of nanosized metal wear debris formed by tribochemical reaction as confirmed by high-resolution TEM and XPS analyses. J Biomed Mater Res Part A 2008;90:1100-10

[15] Ichinose S, Muneta T, Sekiya I, Itoh S, Aoki H, Tagami M. The study of metal ion release and cytotoxicity in Co-Cr-Mo and Ti-Al-V alloy in total knee prosthesis – scanning electron microscopic observation. J Mater Sci: Mater Med 2003;14:79-86.

[16] Urban RM, Jacobs JJ, Tomlinson MJ, Gavrilovic J, Black J, Peoc'h M. Dissemination of wear particles to the liver, spleen, and abdominal lymph nodes of patients with hip or knee replacement. J Bone Joint Surg 2000;82-A:457-77.

[17] Oudadesse H, Irigaray JL, Chassot E. Detection of metallic elements migration around a prosthesis by neutron activation analysis and by the PIXE method. J Trace Microprobe Techniques 2000;18:505-10.

[17a] Chen Z, Wang Z, Wang Q, Cui W, Liu F, Fan W. Changes In early serum metal ion levels and impact on liver, kidney, and immune markers following metal-on-metal total hip arthroplasty. J Arthroplasty 2014;29:612-6.

[18] Doorn PF, Campbell PA, Worrall J, Benya PD, McKellop HA, Amstutz HC. Metal wear particle characterization from metal on metal total hip replacements: Transmission electron microscopy study of periprosthetic tissues and isolated particles. J Biomed Mater Res 1998;42:103-11.

[19] Baslé MF, Bertrand G, Guyetant S, Chappard D, Lesourd M. Migration of metal and polyethylene particles from articular prostheses may generate lymphadenopathy with histiocytosis. J Biomed Mater Res 1996;30:157-64.

[20] Campbell P, Urban RM, Catelas I, Skipor AK, Schmalzried TP. Autopsy analysis thirty years after metal-on-metal total hip replacement. J Bone Joint Surg Incorporated 2003;85:2218-22.

[21] Huber M, Reinisch G, Trettenhahn G, Zweymüller K, Lintner F. Presence of corrosion products and hypersensitivity-associated reactions in periprosthetic tissue after aseptic loosening of total hip replacements with metal bearing surfaces. Acta Biomaterialia 2009;5:172-80.

[22] Al-Hajjar M, Fisher J, Williams S, Tipper JL, Jennings LM. Effect of femoral head size on the wear of metal on metal bearings in total hip replacements under adverse edge-loading conditions. J Biomed Mater Res B: Appl Biomater 2013;101B:213-22.

[23] Catelas I, Bobyn JD, Medley JB, Krygier JJ, Zukor DJ, Petit A, Huk OL. Effects of digestion protocols on the isolation and characterization of metal-metal wear particles. I. Analysis of particle size and shape. J Biomed Mater Res 2001;55:320-9.

[24] Catelas I, Bobyn JD, Medley JB, Krygier JJ, Zukor DJ, Petit A, Huk OL. Effects of digestion protocols on the isolation and characterization of metal-metal wear particles. II. Analysis of ion release and particle composition. J Biomed Mater Res 2001;55:330-7.

[25] Case CP, Langkamer VG, James C, Palmer MR, Kemp AJ, Heap PF, Solomon L. Widespread dissemination of metal debris from implants. J Bone Joint Surg 1994;76-B:701-12.

[26] Passi P, Zadro A, Galassini S, Rossi P, Moschini G. PIXE micro-beam mapping of metals in human peri-implant tissues. J Mater Sci: Mater Med 2002;13:1083-9.

[27] Niinomi M, Nakai M, Hieda J. Development of new metallic alloys for biomedical applications. Acta Biomaterialia 2012;8:3888-903.

[28] Long M, Rack HJ. Titanium alloys in total joint replacement – a materials science perspective. Biomaterials 1998;19:1621-39.

[29] Staiger MP, Pietak AM, Huadmai J. Dias G. Magnesium and its alloys as orthopedic biomaterials: A review. Biomaterials 2006;27:1728-34.

[30] Witte F. The history of biodegradable magnesium implants: A review. Acta Biomaterialia 2010;6:1680-92.

[31] Bansiddhi A, Sargeant TD, Stupp SI, Dunand DC. Porous NiTi for bone implants: A review. Acta Biomaterialia 2008;4:773-82.

[32] Novikova GE. Introduction to corrosion of bioimplants. Protection Metals Phys Chem Surf 2011;47:372-80.

[33] Buford A, Goswami T. Review of wear mechanisms in hip implants: Paper I – General. Mater Design 2004;25:385-93.

[34] Milošev I. Metallic materials for biomedical applications: Laboratory and clinical studies. Pure Appl Chem 2011;83:309-24.

[35] Swiontkowski MF, Agel J, Schwappach J, McNair P, Welch M. Cutaneous metal sensitivity in patients with orthopaedic injuries. J. Orthop Trauma 2001;15:86-9.

[36] Sargeant A, Goswami T. Hip implants - Paper VI - Ion concentrations. Materials Design 2007;28:155-71.

[37] Afolaranmi GA, Tettey J, Meek RMD, Grant MH. Release of chromium from orthopaedic arthroplasties. Open Orthop J 2008;2:10-8.

[38] Marti A. Cobalt-base alloys used in bone surgery. Injury Int J Care Injured 2000;31:S-D18-21.

[39] Okazaki Y, Gotoh E. Comparison of metal release from various metallic biomaterials in vitro. Biomaterials 2005;26:11-21.

[40] Sunderman Jr. FW, Hopfer SM, Swift T, Rezuke WN, Ziebka L, Highman P, Edwards B, et al. Cobalt, chromium, and nickel concentrations in body fluids of patients with porous-coated knee or hip prostheses. J Orthop Res 1989;7:307-15.

[41] Lugowski SJ, Smith DC, McHugh AD, Van Loon JC. Determination of chromium, cobalt and molybdenum in synovial fluid by GFAAS. J Trace Elem Electrolytes Health Dis 1991;5:23-9.

[42] Leopold SS, Berger RA, Patterson L, Skipor AK, Urban RM, Jacobs JJ. Serum titanium level for diagnosis of a failed, metal-backed patellar component. J Arthroplasty 2000;15:938-43.

[43] Lützner J, Dinnebier G, Hartmann A, Günther K-P, Kirschner S. Study rationale and protocol: prospective randomized comparison of metal ion concentrations in the patient's plasma after implantation of coated and uncoated total knee prostheses. BMC Musculoskeletal Dis 2009;10:128-33.

[44] Luetzner J, Krummenauer F, Lengel AM, Ziegler J, Witzleb W-C. Serum metal ion exposure after total knee arthroplasty. Clin Orthop Relat Res 2007;461:136-42.

[45] Friesenbichler J, Maurer-Ertl W, Sadoghi P, Lovse T, Windhager R, Leithner A. Serum metal ion levels after rotating-hinge knee arthroplasty: comparison between a standard device and a megaprosthesis. Int Orthop 2012;36:539-44.

[46] Sarmiento-González A, Marchante-Gayón JM, Tejerina-Lobo JM, Paz-Jiménez J, Sanz-Medel A. High-resolution ICP-MS determination of Ti, V, Cr, Co, Ni, and Mo in human blood and urine of patients implanted with a hip or knee prosthesis. Anal Bioanal Chem 2008;391:2583-9.

[47] Liu T-K, Liu S-H, Chang C-H, Yang R-S. Concentration of metal elements in the blood and urine in the patients with cementless total knee arthroplasty. Tohoku J Exp Med 1998;185:253-62.

[48] Rello L, Lapeña AC, Aramendía M, Belarra MA, Resano M. A dried urine spot test to simultaneously monitor Mo and Ti levels using solid sampling high-resolution continuum source graphite furnace atomic absorption spectrometry. Spectrochim Acta Part B 2013;81:11-9.

[49] Coleman RF, Herrington J, Scales JT. Concentration of wear products in hair, blood, and urine after total hip replacement. Br Med J 1973;1:527-33.

[50] Dumbleton JH, Manley MT. Metal-on-metal total hip replacement. What does the literature say? J Arthroplasty 2005;20:174-88.

[50a] Sampson B, Hart A. Clinical usefulness of blood metal measurements to assess the failure of metal-on-metal hip implants. Ann Clin Biochem 2012;49:118-31.

[51] Pazzaglia UE, Minoia C, Ceciliani L, Riccardi C. Metal determination in organic fluids of patients with stainless steel hip arthroplasty. Acta Orthop Scand 1983;54:574-9.

[52] MacDonald SJ. Can a safe level for metal ions in patients with metal-on-metal total hip arthroplasties be determined?. J Arthroplasty 2004;19:71-7.

[53] Schaffer AW, Pilger A, Engelhardt C, Zweymueller K, Ruediger HW. Increased blood cobalt and chromium after total hip replacement. Clin Toxicol 1999;37:839-44.

[54] Lhotka C, Szekeres T, Steffan I, Zhuber K, Zweymüller K. Four-year study of cobalt and chromium blood levels in patients managed with two different metal-on-metal total hip replacements. J Orthop Res 2003;21:189-95.

[55] Massè A, Bosetti M, Buratti C, Visentin O, Bergadano D, Cannas M. Ion release and chromosomal damage from total hip prostheses with metal-on-metal articulation. Appl Biomaterials Part B 2003;67:750-7.

[56] Ladon D, Doherty A, Newson R, Turner J, Bhamra M, Case CP. Changes in metal levels and chromosome aberrations in the peripheral blood of patients after metal-on-metal hip arthroplasty. J Arthroplasty 2004;19:78-83.

[57] Dunstan E, Sanghrajka AP, Tilley S, Unwin P, Blunn G, Cannon SR, Briggs TWR. Metal ion levels after metal-on-metal proximal femoral replacements. A 30-year follow-up. J Bone Joint Surg 2005;87-B:628-31.

[58] Hart AJ, Hester T, Sinclair K, Powell JJ, Goodship AE, Pele L, Fersht NL, Skinner J. The association between metal ions from hip resurfacing and reduced T-cell counts. J Bone Joint Surg 2006;88-B:449-54.

[59] Antoniou J, Zukor DJ, Mwale F, Minarik W, Petit A, Huk OL. Metal ion levels in the blood of patients after hip resurfacing: A comparison between twenty-eight and thirty-six-millimeter-head metal-on-metal prostheses. J Bone Joint Surg Am 2008;90:142-8.

[60] Daniel J, Ziaee H, Salama A, Pradhan C, McMinn DJW. The effect of the diameter of metal-on-metal bearings on systemic exposure to cobalt and chromium. J Bone Joint Surg 2006;88-B:443-8.

[61] Vendittoli P.-A, Mottard S, Roy AG, Dupont C, Lavigne M. Chromium and cobalt ion release following the Durom high carbon content, forged metal-on-metal surface replacement of the hip. J Bone Joint Surg 2007;89-B:441-8.

[62] Vendittoli P.-A, Roy A, Mottard S, Girard J, Lusignan D, Lavigne M. Metal ion release from bearing wear and corrosion with 28 mm and large-diameter metal-on-metal bearing articulations. A follow-up study. J Bone Joint Surg 2010;92-B:12-9.

[63] Vendittoli P.-A, Riviére C, Roy AG, Barry J, Lusignan D, Lavigne M. Metal-on-metal hip resurfacing compared with 28-mm diameter metal-on-metal total hip replacement. Bone Joint J 2013;95-B:1464-73.

[64] Daniel J, Ziaee H, Pradhan C, Pynsent PB, McMinn DJW. Blood and urine metal ion levels in young and active patients after Birmingham hip resurfacing arthroplasty. J Bone Joint Surg 2007;89-B:169-73.

[64a] Vijaysegaran P, Whitehouse SL, Bijoor M, English H, Crawford RW. Metal ion levels post primary unilateral Exeter total hip arthroplasty. Hip Int 2014;24:144-8.

[65] Daniel J, Ziaee H, Pradhan C, McMinn DJW. Six-year results of a prospective study of metal ion levels in young patients with metal-on-metal hip resurfacings. J Bone Joint Surg 2009;91-B:176-9.

[66] Ziaee H, Daniel J, Datta AK, Blunt S, McMinn DJW. Transplacental transfer of cobalt and chromium in patients with metal-on-metal hip arthroplasty. A controlled study. J Bone Joint Surg 2007;89-B:301-5.

[67] Daniel J, Ziaee H, Pynsent PB, McMinn DJW. The validity of serum levels as a surrogate measure of systemic exposure to metal ions in hip replacement. J Bone Joint Surg 2007;89-B:736-41.

[68] Walter LR, Marel E, Harbury R, Wearne J. Distribution of chromium and cobalt ions in various blood fractions after resurfacing hip arthroplasty. J Arthroplasty 2008;23:814-21.

[69] Smolders JMH, Bisseling P, Hol A, Van der Straeten C, Schreurs BW, van Susante JLC. Metal ion interpretation in resurfacing versus conventional hip arthroplasty and in whole blood versus serum. How should we interpret metal ion data?. Hip Int 2011;21:587-95.

[70] Smolders JMH, Hol A, Rijnberg WJ, van Susante JLC. Metal ion levels and functional results after either resurfacing hip arthroplasty or conventional metal-on-metal hip arthroplasty. Short-term outcome of a randomized controlled trial. Acta Orthop 2011;82:559-66.

[71] Hart AJ, Sabah SA, Bandi AS, Maggiore P, Tarassoli P, Sampson B, Skinner JA. Sensitivity and specificity of blood cobalt and chromium metal ions for predicting failure of metal-on-metal hip replacement. J Bone Joint Surg 2011;93-B:1308-13.

[72] Hart AJ, Skinner JA, Henckel J, Sampson B, Gordon F. Insufficient acetabular version increases blood metal ion levels after meta-on-metal hip resurfacing. Clin Orthop Relat Res 2011;469:2590-7.

[73] Newton AW, Ranganath L, Armstrong C, Peter V, Roberts NB. Differential distribution of cobalt, chromium, and nickel between whole blood, plasma and urine

in patients after metal-on-metal (MoM) hip arthroplasty. J Orthop Res 2012;30:1640-6.

[73a] Parry MC, Eastaugh S, Bannister GC, Learmonth ID, Case CP, Blom AW. Blood levels of cobalt and chromium are inversely correlated to head size after metal-on-metal resurfacing arthroplasty. Hip Int 2013;23:529-34.

[74] Hart AJ, Buddhdev P, Winship P, Faria N, Powell JJ, Skinner JA. Cup inclination angle of greater than 50 degrees increases whole blood concentrations of cobalt and chromium ions after metal-on-metal hip resurfacing. Hip Int 2008;18:212-9.

[75] Davda K, Lali FV, Sampson B, Skinner JA, Hart AJ. An analysis of metal ion levels in th joint fluid of symptomatic patients with metal-on-metal hip replacements. J Bone Joint Surg 2011;93-B:738-45.

[76] Langton DJ, Jameson SS, Joyce TJ, Webb J, Nargol AVF. The effect of component size and orientation on the concentrations of metal ions after resurfacing arthroplasty of the hip. J Bone Joint Surg 2008;90-B:1143-51.

[77] Langton DJ, Jameson SS, Joyce TJ, Hallab NJ, Natu S, Webb J, Nargol AVF. Early failure of metal-on-metal bearings in hip resurfacing and large-diameter total hip replacement. A consequence of excess wear. J Bone Joint Surg 2010;92:38-45.

[78] Desy NM, Bergeron SG, Petit A, Huk OL, Antoniou J. Surgical variables influence metal ion levels after hip resurfacing. Clin Orthop Relat Res 2011;469:1635-41.

[79] Vendittoli P-A, Amzica T, Roy AG, Lusignan D, Girard J, Lavigne M. Metal ion release with large-diameter metal-on-metal hip arthroplasty. J Arthroplasty 2011;26:282-8.

[79a] Robinson PG, Wilkinson AJ, Meek RMD. Metal ion levels and revision rates in metal-on-metal hip resurfacing arthroplasty: a comparative study. Hip Int 2014;24:123-8.

[79b] De Pasquale D, Stea S, Squarzoni S, Bordini B, Amabile M, Catalani S, Apostoli P, Toni a. Metal-on-metal hip prostheses: Correlation between debris in the synovial fluid and levels of cobalt and chromium ions in the bloodstream. Int Orthopaedics 2014;38:469-75.

[80] Omlor GW, Kretzer JP, Reinders J, Streit MR, Bruckner T, Gotterbarm T, Aldinger PR, Merle C. In vivo serum titanium ion levels following modular neck total hip arthroplasty – 10 year results in 67 patients. Acta Biomaterialia 2013;9:6278-82.

[81] Tkaczyk C, Petit A, Antoniou J, Zukor DJ, Tabrizian M, Huk OL. Significance of elevated blood metal ion levels in patients with metal-on-metal prostheses: An evaluation of oxidative stress markers. Open Orthop J 2010;4:221-7.

[82] Case CP, Ellis L, Turner JC, Fairman B. Development of a routine method for the determination of trace metals in whole blood by magnetic sector inductively coupled plasma mass spectrometry with particular relevance to patients with total hip and knee arthroplasty. Clin Chem 2011;47:275-80.

[83] Isaac GH, Siebel T, Oakeshott RD, McLennan-Smith R, Cobb AG, Schmalzried TP, Vail TP. Changes in whole blood metal ion levels following resurfacing: serial measurements in a multi-centre study. Hip Int 2009;19:330-7.

[84] Berstein M, Walsh A, Petit A, Zukor DJ, Antoniou J. Femoral head size does not affect ion values in metal-on-metal total hips. Clin Orthop Relat Res 2011;469:1642-50.

[84a] Bernstein M, Desy NM, Petit A, Zukor DJ, Huk OL, Antoniou J. Long-term follow-up and metal ion trend of patients with metal-on-metal total hip arthroplasty. Int Orthop 2012;36:1807-12.

[85] Pattyn CA, Lauwagie SN, Verdonk RC. Whole blood metal ion concentrations in correlation with activity level in three different metal-on-metal bearings. J Arthroplasty 2011;26:58-64.

[85a] Emmanuel AR, Bergin KM, Kelly GE, McCoy GF, Wozniak AP, Quinlan JF. The effect of acetabular inclination on metal ion levels following metal-on-metal hip arthroplasty. J Arthroplasty 2014;29:186-91.

[85b] Sidaginamale RP, Joyce TJ, Lord JK, Jefferson R, Blain PG, Nargol AVF, Langton DJ. Blood metal ion testing is an effective screening tool to identify poorly performing metal-on-metal bearing surfaces. Bone Joint Res 2013;2:84-95.

[86] Lavigne M, Belzile EL, Roy A, Morin F, Amzica T, Vendittoli P-A. Comparison of whole-blood metal ion levels in four types of metal-on-metal large-diameter femoral head total hip arthroplasty: The potential influence of the adapter sleeve. J Bone Joint Surg 2011;93:128-36.

[86a] Mazoochian F, Schmidutz F, Kiefl J, Fottner A, Michalke B, Schierl R, Thomas P, Jansson V. Levels of Cr, Co, Ni and Mo in erythrocytes, serum and urine after hip resurfacing arthroplasty. Acta Chirurgica Belgica 2013;113:123-8.

[87] Hallab NJ, Jacobs JJ, Skipor A, Black J, Mikecz K, Galante JO. Systematic metal-protein binding associated with total joint replacement arthroplasty. J Biomed Mater Res 2000;49:353-61.

[88] Harding I, Bonomo A, Crawford R, Psychoyios V, Delves T, Murray D, McLardy-Smith P. Serum levels of cobalt and chromium in a complex modular total hip arthroplasty system. J Arthroplasty 2002;17:893-5.

[89] Jacobs JJ, Hallab NJ, Skipor AK, Urban RM. Metal degradation products: A cause for concern in metal-metal bearings? Clin Orthop Rel Res 2003;417:139-47.

[90] Jacobs JJ, Skipor AK, Black J, Urban RM, Galante JO. Release and excretion of metal in patients who have a total hip replacement component made of titanium-base alloy. J Bone Joint Surg Series A 1991;73:1475-86.

[91] Kreibich DN, Moran CG, Delves HT, Owen TD, Pinder IM. Systemic release of cobalt and chromium after uncemented total hip replacement. J Bone Joint Surg 1996;78-B:18-21.

[92] Jacobs JJ, Skipor AK, Patterson LM, Hallab NJ, Paprosky WG, Black J, Galante JO. Metal release in patients who have had a primary total hip arthroplasty: A prospective, controlled, longitudinal study. J Bone Joint Surg Series A 1998;80:1447-58.

[93] Savarino L, Granchi D, Ciapetti G, Cenni E, Pantoli AN, Rotini R, Veronesi CA, et al. Ion release in patients with metal-on-metal hip bearings in total joint replacement: A comparison with metal-on-polyethylene bearings. J Biomed Mater Res 2002;63:467-74.

[94] Savarino L, Granchi D, Ciapetti G, Stea S, Donati ME, Zinghi G, Fontanesi G, et al. Effects of metal ions on white blood cells of patients with failed total joint arthroplasties. J Biomed Mater Res 1999;47:543-50.

[95] Savarino L, Greco M, Cenni E, Cavasinni L, Rotini R, Baldini N, Giunti A. Differences in ion release after ceramic-on-ceramic and metal-on-metal total hip replacement. Medium-term follow-up. J Bone Joint Surg 2006;88-B:472-6.

[96] Savarino L, Padovani G, Ferretti M, Greco M, Cenni E, Perrone G, Greco F, Baldini N, Giunti A. Serum ion levels after ceramic-on-ceramic and metal-on-metal total hip arthroplasty: 8-Year minimum follow-up. J Orthop Res 2008;26:1569-76.

[97] Moroni A, Savarino L, Hoque M, Cadossi M, Baldini N. Do ion levels in hip resurfacing differ from metal-on-metal THA at midterm? Clin Orthop Relat Res 2011;469:180-7.

[98] Savarino L, Cadossi M, Chiarello E, Baldini N, Giannini S. Do ion levels in metal-on-metal hip resurfacing differ from those in metal-on-metal THA at long-term followup? Clin Orthop Relat Res 2013;471:2964-71.

[99] Ball ST, Severns D, Linn M, Meyer RS, Swenson FC. What happens to serum metal ion levels after a metal-on-metal bearing is removed? J Arthroplasty 2013;28 Suppl:53-5.

[100] Skipor AK, Campbell PA, Patterson LM, Anstutz HC, Schmalzried TP, Jacobs JJ. Serum and urine metal levels in patients with metal-on-metal surface arthroplasty. J Mater Sci; Mater Med 2002;13:1227-34.

[101] Brodner W, Bitzan P, Meisinger V, Kaider A, Gottsauner-Wolf F, Kotz R. Elevated serum cobalt with metal-on-metal articulating surfaces. J Bone Joint Surg 1997;79-B:316-21.

[102] Brodner W, Bitzan P, Meisinger V, Kaider A, Gottsauner-Wolf F, Kotz R. Serum cobalt levels after metal-on-metal total hip arthroplasty. J Bone Joint Surg Incorporated 2003;85:2168-73.

[103] Gleizes V, Poupon J, Lazennec JY. Value and limits of determining serum cobalt levels in patients with metal on metal articulating prostheses. Rev Chir Orthop Reparatrice Appar Mot 1999;85:217-24.

[104] Clarke MT, Lee PTH, Arora A, Villar RN. Levels of metal ions after small- and large-diameter metal-on-metal hip arthroplasty. J Bone Joint Surg 2003;85-B:913-7.

[105] Savarino L, Granchi D, Ciapetti G, Cenni E, Greco M, Rotini R, Veronesi CA, et al. Ion release in stable hip arthroplasties using metal-on-metal articulating surfaces: A comparison between short- and medium-term results. J. Biomed Mater Res 2003;66-A:450-6.

[106] Patton MS, Lyon TDB, Ashcroft GP. Levels of systemic metal ions in patients with intramedullary nails. Acta Orthop 2008;79:820-5.

[107] Damie F, Favard L. Metal serum levels in 48 patients bearing a chromium-cobalt total hip arthroplasty with a metal-on-polyethylene combination. Rev Chir Orthop Reparatrice Appar Mot 2004;90:241-8

[108] Maezawa K, Nozawa M, Matsuda K, Yasuma M, Mori K, Enomoto F, Ogawa S, et al. Chronological changes of serum chromium levels after modern metal-on-metal total hip arthroplasty. Acta Orthop Scand 2004;75:422-6.

[109] Maezawa K, Nozawa M, Yuasa T, Aritomi K, Matsuda K, Shitoto K. Seven years of chronological changes of serum chromium levels after metasul metal-on-metal total hip arthroplasty. J Arthroplasty 2010;25:1196-200.

[110] Hasegawa M, Yoshida K, Wakabayashi H, Sudo A. Cobalt and chromium ion release after large-diameter metal-on-metal total hip arthroplasty. J Arthroplasty 2012;27:990-6.

[111] Maezawa K, Nozawa M, Hirose T, Matsuda K, Yasuma M, Shitoto K, Kurosawa H. Cobalt and chromium concentrations in patients with metal-on-metal and other cementless total hip arthroplasty. Arch Orthop Trauma Surg 2002;122:283-7.

[112] Heisel C, Silva M, Skipor AK, Jacobs JJ, Schmalzried TP. The relationship between activity and ions in patients with metal-on-metal bearing hip prostheses. J Bone Joint Surg Incorporated 2005;87:781-7.

[113] Milošev I, Pišot V, Campbell P. Serum levels of cobalt and chromium in patients with Sikomet metal-metal total hip replacements. J Orthop Res 2005;23:526-35.

[114] Iavicoli I, Falcone G, Alessandrelli M, Cresti R, de Santis V, Alimonti A, Carelli G. The release of metals from metal-on-metal surface arthroplasty of the hip. J Trace Elem Med Biol 2006;20:25-31.

[115] Rasquinha VJ, Ranawat CS, Weiskopf J, Rodriguez JA, Skipor AK, Jacobs JJ. Serum metal levels and bearing surfaces in total hip arthroplasty. J Arthroplasty 2006;21:47-52.

[116] Black J, Maitin EC, Gelman H, Morris DM. Serum concentrations of chromium, cobalt and nickel after total hip replacement: A six month study. Biomaterials 1983;4:160-4.

[117] Witzleb W-C, Ziegler J, Krummenauer F, Neumeister V, Guenther K-P. Exposure to chromium, cobalt and molybdenum from metal-on-metal total hip replacement and hip resurfacing arthroplasty. Acta Orthop 2006;77:697-705.

[118] Grübl A, Weissinger M, Brodner W, Gleiss A, Giurea A, Gruber M, Pöll G, et al. Serum aluminium and cobalt levels after ceramic-on-ceramic and metal-on-metal total hip replacement. J Bone Joint Surg 2006;88-B:1003-5.

[119] Sauvé P, Mountney J, Khan T, de Beer J, Higgins B, Grover M. Metal ion levels after metal-on-metal Ring total hip replacement. A 30-year follow-up study. J Bone Joint Surg 2007;89-B:586-90.

[120] Zijlstra WP, van der Veen HC, van den Akker-Scheek I, Zee MJM, Bulstra SK, van Raay JJAM. Acetabular bone density and metal ions after metal-on-metal versus metal-on-polyethylene total hip arthroplasty; short-term results. Hip Int 2014;24:136-43.

[121] Grübl A, Marker M, Brodner W, Giurea A, Heinze G, Meisinger V, Zehetgruber H, Kotz R. Long-term follow-up of metal-on-metal total hip replacement. J Orthop Res 2007;25:841-8.

[122] de Haan R, Pattyn C, Gill HS, Murray DW, Campbell PA, de Smet K. Correlation between inclination of the acetabular component and metal ion levels in metal-on-metal hip resurfacing replacement. J Bone Joint Surg 2008;90-B:1291-7.

[123] Hartmann A, Lützner J, Kirschener S, Witzleb W-C, Günther K-P. Do survival rate and serum ion concentrations 10 years after metal-on-metal hip resurfacing provide evidence for continued use? Clin Orthop Relat Res 2012;470:3118-26.

[124] Langton DJ, Sprowson AP, Joyce TJ, Reed M, Carluke I, Partington P, Nargol VF. Blood metal ion concentrations after hip resurfacing arthroplasty. A comparative study of articular surface replacement and Birmingham hip resurfacing arthroplasties. J Bone Joint Surg 2009;91-B:1287-95.

[125] Lazennec J-Y, Boyer P, Poupon J, Rousseau M-A, Roy C, Ravaud P, Catonné Y. Outcome and serum ion determination up to 11 years after implantation of a cemented metal-on-metal hip prosthesis. Acta Orthop 2009;80:168-73.

[126] Engh Jr. CA, MacDonald SJ, Sritulanondha S, Thompson A, Naudie D, Engh CA. Metal ion levels after metal-on-metal total hip arthroplasty: A randomized trial. Clin Orthop Relat Res 2009;467:101-11.

[127] Van der Straeten C, Grammatopoulos G, Beng HSG, Calistri A, Campbell P, de Smet KA. The interpretation of metal ion levels in unilateral and bilateral hip resurfacing. Clin Orthop Relat Res 2013;471:377-85.

[128] Van der Straeten C, Van Quickenborne B, De Roest B, Calistri A, Victor J, De Smet K. Metal ion levels from well-functioning Birmingham hip resurfacings decline significantly at ten years. Bone Joint J 2013;95-B:1332-8.

[129] Savarino L, Padovani G, Ferretti M, Greco M, Cenni E, Perrone G, Greco F et al. Serum ion levels after ceramic-on-ceramic and metal-on metal total hip arthroplasty: 8-Year minimum follow-up. J Orthop Res 2008;26:1569-76.

[130] Ordóñez YN, Montes-Bayón M, Blanco-González E, Paz-Jiménez J, Tejerina-Lobo JM, Peña-López JM, Sanz-Medel A. Metal release in patients with total hip arthroplasty by DF-ICP-MS and their association to serum proteins. J Anal At Spectrom 2009;24:1037-43.

[131] Hallab NJ, Caicedo M, McAllister K, Skipor A, Amstutz H, Jacobs JJ. Asymptomatic prospective and retrospective cohorts with metal-on-metal hip

arthroplasty indicate acquired lymphocyte reactivity varies with metal ion levels on a group basis. J Orthop Res 2013;31:173-182.

[132] Yoon JP, Le Duff MJ, Johnson AJ, Takamura KM, Ebramzadeh E, Amstutz HC. Contact patch to rim distance predicts metal ion levels in hip resurfacing. Clin Orthop Relat Res 2013;471:1615-21.

[133] Levine BR, Hsu AR, Skipor AK, Hallab NJ, Paprosky WG, Galante JO, Jacobs JJ. Ten-year outcome of serum metal ion levels after primary total hip arthroplasty: A concise follow-up of a previous report. J Bone Joint Surg Am 2013;95:512-8.

[134] Vundelinckx BJ, Verhelst LA, De Schepper J. Taper corrosion in modular hip prostheses analysis of serum metal ions in 19 patients. J Arthroplasty 2013;28:1218-23.

[135] Nelis R, de Waal Malefijt J, Gosens T. Breast milk metal ion levels in a young and active patient with a metal-on-metal hip prosthesis. J Arthroplasty 2013;28:19-22.

[136] Nuevo-Ordóñez Y, Montes-Bayón M, González E, Paz-Aparicio J, Raimundez JD, Tejerina JM, Peña MA, Sanz-Medel A. Titanium release in serum of patients with different bone fixation implants and its interaction with serum biomolecules at physiological levels. Anal Bioanal Chem 2011;401:2747-54.

[137] Granchi D, Savarino L, Ciapetti G, Cenni E, Rotini R, Mieti M, Baldini N, Giunti AJ. Immunological changes in patients with primary osteoarthritis of the hip after total joint replacement. Bone Joint Surg 2003;85-B:758-64.

[138] Johnson AJ, Le Duff MJ, Yoon JP, Al-Hamad M, Amstutz HC. Metal ion levels in total hip arthroplasty versus hip resurfacing. J Arthroplasty 2013;28:1235-7.

[139] Amstutz HC, Campbell PA, Dorey FJ, Johnson AJ, Skipor AK, Jacobs JJ. Do ion concentrations after metal-on-metal hip resurfacing increase over time? A prospective study. J Arthroplasty 2013;28:695-700.

[139a] Le Duff MJ, Johnson AJ, Wassef AJ, Amstutz HC. Does femoral neck to cup impingement affect metal ion levels in hip resurfacing? Clin Orthop Relat Res 2014;472:489-96.

[140] Jacobs JJ, Skipor AK, Doom PF, Campbell P, Schmalzried TP, Black J, Amstutz HC. Cobalt an chromium concentrations in patients with metal on metal total hip replacements. Clin Orthop Relat Res 1996;329 Suppl:S256-63.

[141] Moroni A, Savarino L, Cadossi M, Baldini N, Giannini S. Does ion release differ between hip resurfacing and metal-on-metal THA? Clin Orthop Relat Res 2008;466:700-7.

[142] Heisel C, Streich N, Krachler M, Jakubowitz E, Kretzer JP. Characterization of the running-in period in total hip resurfacing arthroplasty: An *in vivo* and *in vitro* metal ion analysis. J Bone Joint Surg 2008;90:125-33.

[143] Dahlstrand H, Stark A, Anissian L, Hailer NP. Elevated serum concentrations of cobalt, chromium, nickel, and manganese after metal-on-metal alloarthroplasty of the hip: A prospective randomized study. J Arthroplasty 2009;24:837-45.

[144] de Souza RM, Parsons NR, Oni T, Dalton P, Costa M, Krikler S. metal ion levels following resurfacing arthroplasty of the hip. Serial results over a ten-year period. J Bone Joint Surg 2010;92-B:1642-7.

[145] Yang J, Shen B, Zhou ZK, Pei F, Kang PD. Changes in cobalt and chromium levels after metal-on-metal hip resurfacing in young, active Chinese patients. J Arthroplasty 2011;26:65-70.

[146] Lardanchet J.-F, Taviaux J, Arnalsteen D, Gabrion A, Mertl P. One-year prospective comparative study of three large-diameter metal-on-metal total hip prostheses: Serum metal ion levels and clinical outcomes. Orthop Traumattology: Surg Res 2012;98:265-74.

[147] Imanishi T, Hasegawa M, Sudo A. Serum metal ion levels after second-generation metal-on-metal total hip arthroplasty. Arch Orthop Trauma Surg 2010;130:1447-50.

[148] Allan DG, Trammell R, Dyrstad B, Bamhart B, Milbrandt JC. Serum cobalt and chromium elevations following hip resurfacing with the Cormet 2000 device. J Surg Orthop Adv 2007;16:12-8.

[149] Bealé PE, Kim PR, Hamdi A, Fazekas A. A prospective metal ion study of large-head metal-on-metal bearing: A matched-pair analysis of hip resurfacing versus total hip replacement. Orthop Clin North Am 2011;42:251-7.

[150] Garbuz DS, Tanzer M, Greidanus NV, Masri BA, Duncan CP. Metal-on-metal hip resurfacing versus large-diameter head metal-on-metal total hip arthroplasty. Clin Orthop Relat Res 2010;468:318-25.

[151] Brien WW, Salvati EA, Betts F, Bullough P, Wright T, Rimnac C, Buly R, Garvin K. Metal levels in cemented total hip arthroplasty: A comparison of well-fixed and loose implants. Clin Orthop Relat Res 1992;276:66-74.

[152] Nikolaou VS, Petit A, Zukor DJ, Papanastasiou C, Huk OL, Antoniou J. Presence of cobalt and chromium ions in the seminal fluid of young patients with metal-on-metal total hip arthroplasty. J Arthroplasty 2013;28:161-7.

[153] Dobbs HS, Minski MJ. Metal ion release after total hip replacement. Biomaterials 1980;1:193-8.

[154] Chassot E, Irigaray JL, Terver S, Vanneuville G. Contamination by metallic elements released from joint prostheses. Med Eng Phys 2004;26:193-9.

[155] Keegan GM, Learmonth ID, Case CP. Orthopaedic metals and their potential toxicity in the arthroplasty patient. J Bone Joint Surg 2007;89-B:567-73.

[156] Lux F, Zeisler R. Investigations of the corrosive deposition of components of metal implants and of the behavior of biological trace elements in metallosis tissue by means of instrumental multi-element activation analysis. J Radioanal Chem 1974;19:289-97.

[157] Schnabel C, Herpers U, Michel R, Löer F, Buchhorn G, Willert H.-G. Changes of concentrations of the elements Co, Cr, Sb, and Sc in tissues of persons with joint implants. Biol Trace Elem Res 1994;43:389-95.

[157a] Krischak GD, Gebhard F, Mohr W, Krivan V, Ignatius A, Beck A, Wachter NJ, Reuter P, Arand M, Kinzl L, Claes LE. Difference in metallic wear distribution released from commercially pure titanium compared with stainless steel plates. Arch Orthop Trauma Surg 2004;124:104-13.

[158] Agins HJ, Alcock NW, Bansal M, Salvati EA, Wilson PD, Pellicci PM, Bullough PG. Metallic wear in failed titanium-alloy total hip replacements. J Bone Joint Surg 1988;70-A:347-56.

[158a] Lohmann CH, Nuechtem JV, Singh G, Junk-Jantsch S, Schmotzer H, Morlock MM, Pflüger G. Periprosthetic tissue metal content but not serum metal content predict the type of tissue response in failed small-diameter metal-on-metal total hip arthroplasties. J Bone Joint Surg 2013;95:1561-8.

[159] Zeiner M, Zenz P, Lintner F, Schuster E, Schwägerl, Steffan I. Influence on elemental status by hip-endoprostheses. Microchem J 2007;85:145-8.

[160] Rodríguez de la Flor M, Hernández-Vaquero D, Fernández-Carreira JM. Metal presence in hair after metal-on-metal resurfacing arthroplasty. J Orthop Res 2013;31:2015-31.

[161] Vieweg U, Van Roost D, Wolf HK, Schyma CA, Schramm J. Corrosion on an internal spinal fixator system. Spine;1999:24:946-51.

[162] Wang JC, Yu WD, Sandhu HS, Betts F, Bhuta S, Delamarter RB. Metal debris from titanium spinal implants. Spine;1999:899-903.

[162a] Cundy TP, Antoniou G, Sutherland LM, Freeman BJC, Cundy PJ. Serum titanium, niobium, and aluminum levels after instrumented spinal arthrodesis in children. Spine 2013;38:564-70.

[163] Kim Y-J, Kassab F, Berven SH, Zurakowski D, Hresko MT, Emans JB, Kasser JR. Serum levels of nickel and chromium after instrumented posterior spinal arthrodesis. Spine 2005;30:923-6.

[164] del Rio J, Beguiristain J, Duart J. Metal levels in corrosion of spinal implants. Eur Spine J 2007;16:1055-61.

[165] Rylander LS, Milbrandt JC, Armington E, Wilson M, Olysav DJ. Trace metal analysis following locked volar plating for unstable fractures of the distal radius. Iowa Orthop J 2010;30:89-93.

[166] Richardson TD, Pineda SJ, Strenge KB, Van Fleet TA, MacGregor M, Milbrandt JC, Espinosa JA, Freitag P. Serum titanium levels after instrumental spinal arthrodesis. Spine;33:792-6.

[167] Kasai Y, Iida R, Uchida A. Metal concentrations in the serum and hair of patients with titanium alloy spinal implants. Spine;28:1320-6.

[167a] Cundy TP, Kirby CP. Serum metal levels after minimally invasive repair of pectus excavatum. J Pediatric Surg 2012;47:1506-11.

[167b] Gornet MF, Burkus JK, Harper ML, Chan FW, Skipor AK, Jacobs JJ. Prospective study on serum metal levels in patients with metal-on-metal lumbar disc arthroplasty. Eur Spine J 2013;22:741-6.

[168] Schliephake H, Scharnweber D. Chemical and biological functionalization of titanium for dental implants. J Mater Chem 2008;18:2404-14.

[168a] Wataha JC. Biocompatibility of dental casting alloys: A review. J Prosthetic Dent 2000;83:223-34.

[169] Conrad HJ, Seong WJ, Pesun IJ. Current ceramic materials and systems with clinical recommendations: a systematic review. J Prosthet Dent 2007;98:389-404.

[170] Schmalz G, Garhammer P. Biological interactions of dental cast alloys with oral tissues. Dent Mat 2002;18:396-406.

[171] Geurtsen W. Biocompatibility of dental casting alloys. Crit Rev Oral Biol Med 2002;13:71-84.

[172] House K, Sernetz F, Dymock D, Sandy JR, Ireland AJ. Corrosion of orthodontic appliances-should we care? Am J Orthod Dentofacial Orthop 2008;133:584-92.

[173] Adya N, Alam M, Ravindranath T, Mubeen A, Saluja B. Corrosion in titanium dental implants: Literature review. J Indian Prosthod Soc 2005;5:126-131.

[174] Siddiqi A, Payne AGT, De Silva RK, Duncan WJ. Titanium allergy: Could it affect dental implant integration? Clin Oral Impl Res 2011;22:673-80.

[175] Brownawell AM, Berent S, Brent RL, Bruckner JV, Doull J, Gershwin EM, Hood RD, et al. The potential adverse health effects of dental amalgam. Toxicol Rev 2005;24:1-10.

[176] Roberts HW, Charlton DG. The release of mercury from amalgam restorations and its health effects: A review. Operative Dent 2009;34:605-14.

[177] Rathore M, Singh A, Pant VA. The dental amalgam toxicity fear: A myth or actuality. Toxicol Int 2012;19:81-88.

[177a] Brune D. Metal release from dental biomaterials. Biomaterials 1986;7:163-75.

[177b] Mikulewicz M, Chojnacka K. Trace metal release from orthodontic appliances by in vivo studies: A systematic literature review. Biol Trace Elem Res 2010;137:127-38.

[177c] Macedo de Menezes L, Quintão CCA. The release of ions from metallic orthodontic appliances. Semin Orthod 2010;16:282-92.

[178] Green NT, Machtei EE, Horwitz J, Peled M. Fracture of dental implants: Literature review and report of a case. Impl Dent 2002;11:137-43.

[179] Amini F, Jafari A, Amini P, Sepasi S. Metal Ion release from fixed orthodontic appliances – an in vivo study. Eur J Orthod 2012;34:126-30.

[180] Eliades T, Athanasiou AE. In vivo aging of orthodontic alloys: Implications for corrosion potential, nickel release, and biocompatibility. Angle Orthod 2002;72:222-37.

[181] Eliades T, Trapalis C, Eliades G, Katsavrias E. Salivary metal levels of orthodontic patients: A novel methodological and analytical approach. Eur J Orthod 2003;25:103-6.

[182] Amini F, Farahani AB, Jafari A, Rabbani M. In vivo study of metal content of oral mucosa cells in patients with and without fixed orthodontic appliances. Orthod Craniofac Res 2008;11:51-6.

[182a] Martín-Cameán A, Jos A, Calleja A, Gil F, Iglesias A, Solano E, Cameán AM. Validation of a method to quantify titanium, vanadium and zirconium in oral mucosa cells by inductively coupled plasma-mass spectrometry (ICP-MS). Talanta 2014;118:238-44.

[182b] Martín-Cameán A, Jos A, Calleja A, Gil F, Iglesias-Lineres A, Solano E, Cameán AM. Development and validation of an inductively coupled plasma mass spectrometry (ICP-MS) method for the determination of cobalt, chromium, copper and nickel in oral mucosa cells. Microchem J 2014;114:73-9.

[183] Kerosuo H, Moe G, Hensten-Pettersen A. Salivary nickel and chromium in subjects with different types of fixed orthodontic appliances. Am J Orthod Dentofac Orthop 1997;111:595-8.

[184] Kocadereli I, Ataç A, Kale S, Özer D. Salivary nickel and chromium in patients with fixed orthodontic appliances. Angle Orthod 2000;70:431-4.

[185] Ağaoğlu G, Arun T, Izgü B, Yarat A. Nickel and chromium levels in the saliva and serum of patients with fixed orthodontic appliances. Angle Orthod 2001;71:375-9.

[186] Faccioni F, Franceschetti P, Cerpelloni M, Francasso ME. In vivo study on metal release from fixed orthodontic appliances and DNA damage in oral mucosa cells. Am J Orthod Dentofac Orthop 2003;124:687-93.

[186a] Fernández-Miñano E, Ortiz C, Vicente A, Calvo JL, Ortiz A. Metallic ion content and damage to the DNA in oral mucosa cells of children with fixed orthodontic appliances. Biometals 2011;24:935-41.

[186b] Natarajan M, Padmanabhan S, Chitharanjan A, Narasimhan M. Evaluation of the genotoxic effects of fixed appliances on oral mucosal cells and the relationship to nickel and chromium concentrations: An in-vivo study. Am J Orthod Dentofacial Orthop 2011;140:383-8.

[186c] Hafez HS, Selim EMN, Eid FHK, Tawfik WA, Al-Ashkar EA, Mostafa YA. Cytotoxicity, genotoxicity, and metal release in patients with fixed orthodontic appliances: A longitudinal in-vivo study. Am J Orthod Dentofacial Orthop 2011;140:298-308.

[187] Fors R, Persson M. Nickel in dental plaque and saliva in patients with and without orthodontic appliances. Eur J Orthod 2006;28:292-7.

[188] Sarmiento-González A, Marchante-Gayón JM, Tejerina-Lobo JM, Paz-Jiménez J, Sanz-Medel A. ICP-MS multielemental determination of metals potentially released from dental implants and articular prostheses in human biological fluids. Anal Bioanal Chem 2005;382:1001-9.

[189] Singh DP, Sehgal V, Pradhan KL, Chandna A, Gupta R. Estimation of nickel and chromium in saliva of patients with fixed orthodontic appliances. World J Orthod 2008;9:196-202.

[189a] Petoumenou E, Kislyuk M, Hoederath H, Keilig L, Bourauel C, Jäger A. Corrosion susceptibility and nickel release of nickel titanium wires during clinical application. J Orofacial Orthop 2008;6:411-23.

[190] Petoumenou E, Arndt M, Keilig L, Reimann S, Hoederath H, Eliades T, Jäger A, Bourauel C. Nickel concentration in the saliva of patients with nickel-titanium orthodontic appliances. Am J Orthod Dentofac Orthop 2009;135:59-65.

[190a] Sahoo N, Kailasam V, Padmanabhan S, Chitharanjan AB. In-vivo evaluation of salivary nickel and chromium levels in conventional and self-ligating brackets. Am J Orthod Dentofacial Orthop 2011;140:340-5.

[191] Matos de Souza R, Macedo de Menezes L. Nickel, chromium and iron levels in the saliva of patients with simulated fixed orthodontic appliances. Angle Orthod 2008;78:345-50.

[192] Elshahawy W, Ajlouni R, James W, Abdellatif H, Watanabe I. Elemental ion release from fixed restorative materials into patient saliva. J Oral Rehabil 2013;40:381-8.

[192a] Menezes LM, Quintão CA, Bolognese AM. Urinary excretion levels of nickel in orthodontic patients. Am J Orthod Dentofacial Orthop 2007;131:635-8.

[193] Begerow J, Neuendorf J, Turfeld M, Raab W, Dunemann L. Long-term urinary platinum, palladium, and gold excretion of patients after insertion of noble-metal dental alloys. Biomarkers 1999;4:27-36.

[194] Berglund F, Carlmark B. Titanium, sinusitis, and the yellow nail syndrome. Biol Trace Elem Res 2011;143:1-7.

[195] Meningaud J.-P, Poupon J, Bertrand J.-Ch, Chenevier C, Galliot-Guilley M, Guilbert F. Dynamic study about metal release from titanium miniplates in maxillofacial surgery. Int J Oral Maxillofac Surg 2001;30:185-8.

[196] Zaffe D, Bertoldi C, Consolo U. Element release from titanium devices used in oral and maxillofacial surgery. Biomaterials 2003;24:1093-9.

[197] Templeton DM, Ariese F, Cornelis R, Danielsson L-G, Muntau H, Van Leeuwen HP, Łobiński R. Guidelines for terms related to chemical speciation and fractionation of elements. Definitions, structural aspects, and methodological approaches. Pure Appl Chem 2000;72:1453-70.

[198] Merritt K, Brown SA. Release of hexavalent chromium from corrosion of stainless steel and cobalt – chromium alloys. J Biomed Mater Res 1995;29:627-33.

[199] Afolaranmi GA, Tettey JNA, Murray HM, Meek RMD, Grant MH. The effect of anticoagulants on the distribution of chromium VI in blood fractions. J Arthroplasty 2010;25:118-20.

[199a] Hart AJ, Quinn PD, Sampson B, Sandison A, Atkinson KD, Skinner JA, Powell JJ, Mosselmans JFW. The chemical form of metallic debris in tissues

surrounding metal-on-metal hips with unexplained failure. Acta Biomaterialia 2010;6:4439-46.

[200] Ektessabi A, Shikine S, Kitamura N, Rokkum M, Johansson C. Distribution and chemical states of iron and chromium released from orthopedic implants into human tissues. X-Ray Spectrom 2011;30:44-8.

[201] Sample preparation for trace element analysis (Eds. Mester Z, Sturgeon R). Amsterdam: Elsevier; 2003.

[202] Rodushkin I, Engström E, Baxter DC. Sources of contamination and remedial strategies in the multi-elemental trace analysis laboratory. Anal Bioanal Chem 2010;396:365-77.

[203] Sánchez Misiego A, García-Moncó Carra RM, Ambel Carracedo MP. Trace-element analysis in biological fluids: Evaluation of contamination in samples collected with metallic needles. Anal Chim Acta 2003;494:167-76.

[204] Cornelis R, Heinzow B, Herber RFM, Molin Christensen J, Poulsen OM, Sabbioni E, Templeton DM, et al. Sample collection guidelines for trace elements in blood and urine. J Trace Elements Med Biol 1996;10:103-27.

[205] Anand VD, White JM, Nino HV. Some aspects of specimen collection and stability in trace element analysis of body fluids. Clin Chem 1975;21:595-602.

[206] Musteata FM. Recent progress in in-vivo sampling and analysis. Trends Anal Chem 2013;45:154-68.

[207] Caroli S, Alimonti A, Coni E, Petrucci F, Senofonte O, Violante N. The assessment of reference values for elements in human biological tissues and fluids: A systematic review. Crit Rev Anal Chem 1994;24:363-98.

[208] Roelandts I. Biological and environmental reference materials: Update 1996. Spectrochim Acta Part B 1997;52:1073-86.

[209] Sariego Muñiz S, Fernández-Martin JL, Marchante-Gayón JM, Garcia Alonso JI, Cannata-Andía JB, Sanz-Medel A. Reference values for trace and ultratrace elements in human serum determined by double-focusing ICP-MS. Biol Trace Elem Res 2001;82:259-72.

[210] Cornelis R, Fuentes-Arderiu X, Bruunshuus I, Templeton D. International Union of Pure and Applied Chemistry. Clinical Chemistry Division, Commission on Toxicology, Commission on Nomenclature, Properties and Units and International Federation of Clinical Chemistry, Scientific Division Committee on Nomenclature, Properties and Units. Properties and Units in the Clinical Laboratory Sciences IX. Properties and Units in Trace Elements. Clin Chim Acta 1997;268:S75-S89.

[211] MacDonald SJ, Brodner W, Jacobs JJ. A consensus paper on metal ions in metal-on-metal hip arthroplasties. J Arthroplasty 2004;19:12-6.

[212] Jacobs JJ, Skipor AK, Campbell PA, Hallab NJ, Urban RM, Amstutz HC. Can metal levels be used to monitor metal-on-metal hip arthroplasties. J Arthroplasty 2004;19:59-65.

[213] Jacobs JJ, Hallab NJ, Skipor AK, Urban RM. Metal degradation products. A cause for concern in metal-metal bearings? Clin Orthop Relat Res 2003;417:139-47.

[214] Delaunay C, Petit I, Learmonth ID, Oger P, Vendittoli PA. Metal-on-metal bearings total hip arthroplasty: The cobalt and chromium ions release concern. Orthop Traumat: Surg Res 2010;96:894-904.

[214a] Matusiewicz H. Potential release of in vivo trace metals from metallic medical implants in the human body: From ions to nanoparticles - A systematic analytical review. Acta Biomaterialia 2014;10:2379-403

[215] Jantzen C, Jørgensen HL, Duus BR, Sporring SL, Lauritzen JB. Chromium and cobalt ion concentrations in blood and serum following various types of metal-on-metal hip arthroplasties. Acta Orthop 2013;84:229-36.

[216] Hart AJ, Muirhead-Allwood S, Porter M, Matthies A, Ilo K, Maggiore P, Underwood R, *etc.* Which factors determine the wear rate of large-diameter metal-on-metal hip replacements? J Bone Joint Surg Am 2013;95:678-85.

[216a] Kretzer JP, Van der Straeten C, Sonntag R, Müller U, Streit M, Moradi B, Jäger S, Reinders J. metal ion concentrations in patients with metal-metal bearings in prostheses. Orthopade 2013;42:622-8.

[216b] Parsons PJ, Barbosa Jr. F. Atomic spectrometry and trends in clinical laboratory medicine. Spectrochim Acta Part B 2007;62:992-1003.

[217] Barry J, Lavigne M, Vendittoli P-A. Evaluation of the method for analyzing chromium, cobalt and titanium ion levels in the blood following hip replacement with a metal-on-metal prosthesis. J Anal Toxicol 2013;37:90-6.

[217a] Moyano F, Verni E, Tamashiro H, Digenaro S, Martinez LD, Gil RA. Single-step procedure for trace element determination in synovial fluid by dynamic reaction cell-inductively coupled plasma mass spectrometry. Microchem J 2014;112:17-24.

[217b] Price D, Abou-Shakra F, Jung L, Sieniawska C, Thomas R. The benefits of io-molecule chemistry for the determination of titanium in whole blood and serum using quadrupole-based collision-reaction cell ICP-MS technology. Spectroscopy 2013;11:28-34.

[218] Balcaen L, Bolea-Fernandez E, Resano M, Vanhaecke F. Accurate determination of ultra-trace levels of Ti in blood serum using ICP-MS/MS. Anal Chim Acta 2014;809:1-8.

[218a] Becker JS, Matusch A, Wu B. Bioimaging mass spectrometry of trace elements – recent advance and applications of LA-ICP-MS: A review. Anal Chim Acta 2014;835:1-18.

[219] Chen J, Dong X, Zhao J, Tang G. *In vivo* acute toxicity of titanium dioxide nanoparticles to mice after intraperitioneal injection. J Appl Toxicol 2009;29:330-7.

[220] Nazarenko Y, Zhen H, Han T, Lioy P, Mainelis G. Potential for inhalation exposure to engineered nanoparticles from nanotechnology-based cosmetic powders. Environ Health Perspect 2012;120:885-92.

[221] Zachariadis GA, Sahanidou E. Analytical performance of a fast multi-element method for titanium and trace elements determination in cosmetics and pharmaceuticals by ICP-AES. Cent Eur J Chem 2011;9:213-7.

[221a] Krystek P, Tentschert J, Nia Y, Trouiller B, Noël L, Goetz ME, Papin A, Luch A, Guérin T, de Jong WH. Method development and inter-laboratory comparison about the determination of titanium from titanium dioxide nanoparticles in tissues by inductively coupled plasma mass spectrometry. Anal Bioanal Chem 2014;406:3853-61.

[222] Avula B, Wang Y-H, Duzgoren-Aydin NS, Khan IA. Inorganic elemental compositions of commercial multivitamin/mineral dietary supplements: Application of collision/reaction cell inductively coupled – mass spectroscopy. Food Chem 2011;127:54-62.

[222a] Capelli C, Foppiano D, Venturelli G, Carlini E, Magi E, Ianni C. Determination of arsenic, cadmium, cobalt, chromium, nickel, and lead in cosmetic face-powders: Optimization of extraction and validation. Anal Lett 2014;47:1201-14.

[223] Lomer MCE, Thompson RPH, Commisso J, Keen CL, Powell JJ. Determination of titanium dioxide in foods using inductively coupled plasma optical emission spectrometry. Analyst 2000;125:2339-43.

[223a] López-Heras I, Madrid Y, Cámara C. Prospects and difficulties in TiO_2 nanoparticles analysis in cosmetic and food products using asymmetrical flow-field fractionation hyphenated to inductively coupled plasma mass spectrometry. Talanta 2014;124:71-8.

[224] Smolders JMH, Hol A, van Susante JLC. Metal ion trend may be more predictive for malfunctioning metal-on-metal implants than a single measurement. Hip Int 2013;23:434-40.

[224a] Griffin WL, Metal ion levels: How can they help us? J Arthroplasty 2014;29:659-60.

[225] Macnair RD, Wynn-Jones H, Wimhurst JA, Toms A, Cahir J. Metal ion levels not sufficient as a screening measure for adverse reactions in metal-on-metal hip arthroplasties. J Arthroplasty 2013;28:78-83.

Table 1. Concentration of trace metals in clinical fluids of patients following ion release from different models of MoM knee prostheses.

Type of implant	Body fluid analyzed	Metal ion(s) measured	Concentration (µg L⁻¹)		Detection mode	Sample pre-treatment	Refs.
			MoM	Control			
Cobalt alloy	Synovial fluid, blood	Co, Cr, Mo			GF AAS;L	Three methods were employed to analyze each sample: (1) dilution of the samples with Triton X-100, (2) microwave-assisted decomposition of the samples and (3) classic nitric-perchloric acid decomposition of the samples.	[41]
Co-Cr implant (Miller-Galante TKA)	Serum	Ti	536.8	2.4	GF AAS;L	Serum samples were obtained by allowing room-temperature coagulation and performing centrifugation at 1850g gravity.	[42]
Cr-Co-Mb alloy (Aesculap)	Plasma	Cr Co, Mo, Ni	201	0.92	GF AAS;L	Plasma was separated by centrifugation at 2000 g for 10 min. Samples were diluted 1:2 in buffer, 0.2% Triton X-100, 0.2% Antifoam B; Cr and Co: additional 0.8% Pd-matrix-modifier, 0.3% Mg-matrix-modifier.	[43]
Co-Cr-Mo alloy (Limb Preservation System)	Serum	Co Cr Mo	7.5 2.98 0.4	0.3 0.33 0.4	ET AAS;L	All specimens were centrifuged at 4000 rpm within 2 h and stored at -10°C until analysis.	[44]
Ti6Al4V (Interax-Stryker Howmedica Osteonis)	Blood	Ti V Cr Co Ni Mo	0.590-2.306 0.042-0.181 0.130-0.420 0.224-0.565 0.505-1.199 0.407-1.426	0.300-0.890 0.026-0.280 0.200-0.470 0.040-0.200 0.300-0.770 0.300-1.500	(HR)-ICP-MS;L	Whole blood was digested in a mixture of HNO₃(2 mL) and H₂O₂(1 mL) in closed PTFE vessels. For urine, 1%(v/v) of HNO₃ was added to the samples.	[46]
	Urine	Ti V Cr Co Ni Mo	0.004-0.013 0.002-0.004 0.004-0.009 0.004-0.013 0.015-0.023 0.260-0.427	0.100-0.180 0.008-0.120 0.040-0.300 0.040-0.810 0.240-2.700 12.00-110.0			
Ti6Al4V (3MG-II, Zimmer Co.)	Blood	Co Cr Ti	116.1 108.1 319.6	0.8 5.8 9.3	GF AAS;L	Determination of metals in whole blood was achieved by applying 1g of ammonium phosphate and 1 mL of Triton X-100 diluted with water (1:10). Metal levels in 24 h urine samples (10 mL) were determined; the debris was removed by centrifugation.	[47]
Cr-Co-Mo alloy (8PCA,5 Osteonics Co.)	Urine	Co Cr Ti	0.8 1.1 10.1	0.7 0.6 5.2			

| Metallic implants Not stated | Urine | Mo Ti | 100 650 | Not reported | HR-CS GF AAS;S | Direct determination of trace metals in urine samples deposited on clinical filter papers. | [48] |

L=liquid, S=solid

Table 2. Concentration of trace metals in clinical fluids of patients following ion release from different models of MoM hip prostheses.

Type of implant	Body fluid analyzed	Metal ion(s) measured	Concentration ($\mu g\ L^{-1}$) MoM	Control	Detection mode	Sample pre-treatment	Refs.
Sikomet-SM21	Blood	Co Cr	1.5 2.2	1.1 1.8	GF AAS;L	Blood samples were acid digested (HNO_3,H_2O_2). An $Mg(NO_3)_2$ solution was used as the matrix modifier. Urinary specimens were acidified.	[53]
	Urine	Co Cr	5.5 2.7	0.4 0.3			
Metasul-SM21 THA	Whole Blood	Co Cr	16.95 ngg^{-1} 25.62 ngg^{-1}	0.7 ngg^{-1} 0.21 ngg^{-1}	GF AAS;L	The samples were freeze-dried and totally mineralized using HNO_3 as an ashing reagent.	[54]
Metasul	Blood	Co Cr Mo Ni	1.43-2.32 1.57-1.70 1.09-1.71 2.01-3.71	0.96 0.75 0.80 1.40	GF AAS;L	The blood samples were analyzed after dilution 1:1 with Triton X-100 0.1% and subsequent direct analysis. The urine samples were directly analyzed without dilution.	[55]
	Urine	Co Cr Mo Ni	2.32-10.07 2.10-2.91 4.38-10.49 2.38-3.88	1.04 0.41 4.25 1.36			
Co-Cr-Mo alloy (manufactured)	Whole blood	Co Cr Ti V	35.5 ngg^{-1} 2.7 ngg^{-1} 1.65 ngg^{-1} 0.86 ngg^{-1}	1 1 1 1	(HR)-ICP-MS;L	Blood samples were diluted with water, and 0.1% Triton X-100. Urine samples were diluted with water alone.	[57]
	Urine	Co Cr Ti V	205 ngg^{-1} 3.4 ngg^{-1} 0.9 ngg^{-1} 1.67 ngg^{-1}	1 1 1 1			
Birmingham hip resurfacing	Blood	Co Cr	4.18 1.78	0.5 2.3	ICP-MS;L	Blood samples were transferred to EDTA tubes.	[58]
Metasul, THA	Whole	Co	1.8-2.5	1.75	(HR)-ICP-	Blood samples were diluted 10-fold with a solution of Triton X-100(10 $molL^{-1}$), 0.0002	[59]

Device/Material	Sample	Element	Value	LOD	Method (unit)	Sample preparation	Ref
Ultamet, THA ASR, HRS	Blood	Cr Mo	0.25-0.50 1.30-1.65	0.05 1.3	MS;L	molL^{-1} EDTA, and 0.01 molL^{-1} ammonium hydroxide in water.	
Metasul, THR, BHR, HRS	Whole blood Urine	Co Cr Co Cr	1.3-1.7 1.7-2.4 11.6-14.2 3.7-7.0	Not reported	ICP-MS;L	A reagent containing 0.01 M ammonia, 0.0002 M EDTA and 1% Triton X-100 was used as the stock diluent for whole blood samples. Urine samples were diluted in 1% HNO$_3$	[60]
BHR	Whole blood Serum Plasma	Co Cr Co Cr Co Cr	19.62 nmolL^{-1} 23.13 nmolL^{-1} 23.28 nmolL^{-1} 53.40 nmolL^{-1} 23.97 nmolL^{-1} 57.00 nmolL^{-1}	Not detected	ICP-MS;L GF AAS;L	Blood samples were heparinized as an anticoagulant.	[68]
Co-Cr alloy	Whole blood Plasma Urine	Co Cr Ni Co Cr Ni Co Cr Ni	45.9 nmolL^{-1} 40.39 nmolL^{-1} 51.09 nmolL^{-1} 39.1 nmolL^{-1} 53.81 nmolL^{-1} 41.01 nmolL^{-1} 334.01 nmolL^{-1} 97.31 nmolL^{-1} 41.21 nmolL^{-1}	10 nmolL^{-1} 40 nmolL^{-1} 40 nmolL^{-1} 10 nmoll^{-1} 10 nmoll^{-1} 20 nmol^{-1} Not detected	ICP-MS;L	Blood samples were collected in contamination-free EDTA blood tubes.	[73]
Hip Hip, Knee Prostheses	Whole blood Whole blood	Co Cr Co Cr Mo Ni	1.6-4.45 1.88-4.3 0.17 0.22 0.62 0.99	Not reported Not reported	ICP-MS;L (HR)-ICP-MS;L	10 mL of blood was transferred to an EDTA tube, and samples were stored at -20 C^o Blood samples were diluted 10-fold with a diluent consisting of 0.01 M ammonium hydroxide, 0.0002 M EDTA, and 10 mLL^{-1} Triton X-100 in water	[74] [82]
Metasul, THA	Serum	Co Cr Mo	1.33 1.72 0.62	0.24 0.25 0.27	GF AAS;L	Serum was separated from peripheral blood by centrifugation at 400×g, 10 min, 4°C, dilution with 0.1% HNO$_3$, 0.05% Triton X-100 and Mg(NO$_3$)$_2$	[93]
Metasul, THA	Serum	Al Co Cr	1.36-16.31 0.08-7.31 0.06-8.60	8.28 0.40 0.53	GF AAS;L	Serum was separated by centrifugation at 400g for 10 min at 4 C^o, and frozen at -70 C^o until analysis	[95]

Implant	Sample	Element			Method	Sample preparation	Ref
Conserve plus, HRS	Serum	Ti Co Cr	2.91-11.60 1.07-1.26 1.88-2.02	5.13 0.17 0.08	GF AAS;L	Blood was allowed to clot naturally, was centrifuged and separated into cell and serum fractions and stored at -80 C°.	[100]
Alloclasic Metasul	Urine	Cr	1.40-2.21	0.16	GF AAS;L	Not noted	[101]
	Serum	Co	1.1-16.6				
Ultima, THA	Serum	Co Cr	38 nmolL^{-1} 53 nmolL^{-1}	5 nmolL^{-1} 5 nmolL^{-1}	ICP-MS;L	Blood was centrifuged at 3000 rpm for 10 min and plasma was frozen at -80°C.	[104]
Metasul	Serum	Cr	1.05-1.61	Not reported	GF AAS;L	Not noted	[108]
Sikomet-SM21,THA	Serum	Co Cr	0.26 1.31	0.10 0.30	AdSV;L	After blood centrifugation, serum was transferred to vials. The serum was acid digested by a combination of H_2SO_4, HNO_3, and H_2O_2 in a 10 mL Kjeldhal flask.	[113]
Conserve Plus	Serum	Co Cr	14.0 2.88	3.46 1.47	ICP-MS;L	Serum and urine samples were diluted 1:5 (v/v) with water prior to analysis	[114]
	Urine	Co Cr	5.46 2.00	0.27 0.82			
M/M, THA	Serum	Co Cr Ti	2.35 3.48 1.87	Not reported	GF AAS;L	Blood was centrifuged for 25 min at 4000 rpm and then was separated into serum	[115]
Metasul, THA	Serum	Co Cr Mo	2.17 5.12 2.11	0.25 0.25 2.11	GF AAS;L	Serum was separated by centrifugation at 2000 g for 10 min. The samples were diluted 1:2 in buffer (1% HNO_3), 0.2% Triton X-100, 0.2% Antifoam B; Cr and Co: additional 0.8% Pd-matrix modifier, 0.3% Mg-matrix modifier	[117]
Metasul (Zimmer)	Serum	Al Co	1.9 1.4	1.9 0.15	GF AAS;L	Not noted	[118]
Ring THR	Serum	Co Cr	≈4.09 nmolL^{-1} ≈8.37 nmolL^{-1}	6 nmolL^{-1} 20 nmolL^{-1}	ICP-MS;L	Blood samples was centrifuged at 2500 rpm for 10 min, and the plasma separated into the tubes	[119]
BHR, HRS	Whole blood Serum	Co Cr	3.8-4.2 4.2-8.4	<0.1 <0.2	ICP-MS;L	Whole blood specimens were drawn into a lithium heparin Vacutainer tubes and stored at -18 C°. In order to obtain serum, blood was drawn into a plain Vacutainer tube and centrifuged at 5000 rpm for 10 min and the supernatant stored frozen at -18°C	[120]
50 Conserve Plus, BHR	Blood Serum	Co Cr	9.8 9.7	Not reported	ICP-MS;L	Not noted	[122]

Implant	Sample	Element	Value	Detection limit	Method	Sample preparation	Ref.
Metasul, THR	Serum	Co Cr Ti	1.30-1.69 1.42-2.18 0.70	Not reported	GF AAS;L ICP-OES;L	Not noted	[125]
AML Prodigy	Erythrocyte	Co Cr Ti	0.249 0.609 0.391	0.11 0.90 0.25	ICP-MS;L	Not noted	[126]
	Serum	Co Cr Ti	0.252 0.904 0.85	0.14 0.21 0.20			
	Urine	Co Cr Ti	0.809 0.819 0.192	0.41 0.24 0.09			
Conserve Plus,HRA	Serum	Co Cr	4.0-5.0 4.6-7.4	Not reported	ICP-MS;L	Not noted	[127]
M/M	Whole blood	Co Cr Mo Mn Ti	0.16-0.28 0.56-0.63 0.4-58.6 10.6-11.1 2.1-3.5	0.20 0.19 0.37 6.5 0.5	ICP-MS;L	For whole blood analysis, aliquots of 1 mL were placed into the digestion Teflon vessels and mixed with 6 mL of nitric acid diluted (1:3 with water and 1 mL of hydrogen peroxide). After digestion, the samples were transferred into bottles and finally diluted 1:20 with water	[130]
	Urine	Co Cr Mo Mn Ti	4.4 2.1 37 0.4 0.8	1.0 0.9 33 0.1 0.2			
Conserve Plus	Serum	Co Cr Ni	1.2-3.4 1.5-5.4	<1 <1	ICP-MS;L	Not noted	[131]
Conserve Plus	Serum	Co Cr	1.13 1.49	Not reported	ICP-MS;L	The samples were allowed to clot naturally before being centrifuged (serum)	[132]
T2 Nails Synthes	Serum	Ti	4.18	0.25	ICP-MS;L	Blood samples were separated into serum and RBCs by centrifugation at 3000 g for 15 min	[136]
THR	Serum	Al Co Cr	7.07 1.22 1.54	6.92 0.42 0.29	GF AAS;L	Blood was centrifuged to separate serum	[137]

Device	Sample	Element			Method	Sample preparation	Ref
		Mo	0.65	0.42			
		Ti	3.19	2.91			
Conserve Plus	Serum	Co	1.06-2.80	<0.3	GF AAS;L	Blood was allowed to clot naturally, centrifuged, separated into serum fractions and stored at -80 C° until analysis	[139]
		Cr	1.58-5.80	0.07			
		Co	1.58-5.80		(HR)-ICP-MS;L		
		Cr	1.06-2.80				
Hip	Serum	Co	1.33-1.40	0.08-0.86	GF AAS;L	Serum was separated by centrifugation at 400 x g for 10 min at 4 C°. Specimens were diluted with 0.1 % vol HNO_3 and 0.05% vol Triton X-100 and analyzed. For Cr and Co analysis, magnesium nitrate was added as a matrix modifier	[141]
		Cr	1.73-2.30	0.06-0.67			
		Mo	0.90-0.96	<LOD			
		Ni	0.70	0.10-1.69			
Hip	Serum	Co	0.7	0.3	ICP-MS;L	Not noted	[147]
		Cr	0.6	0.3			
Hip	Serum	Co	5.09	0.11	(HR)-ICP-MS; L	Blood was allowed to clot for 20 min then centrifuged for 15 min. The serum was then transferred into a tube	[150]
		Cr	2.14	0.20			
DePuy ASR System	Serum	Co	25.81	Not reported	ICP-MS;L	Blood samples were collected in special tubes with no anti-coagulant. Serum was obtained by centrifugation. Hair samples were reconditioned with an ultrasonic bath, then, samples were acid digested in a microwave oven.	[160]
		Cr	23.08				
	Urine	Co	205.62				
		Cr	42.83				
	Hair	Co	147.40				
		Cr	185.32				
		Mo	39.31				

L=liquid, S=solid

Table 3. Concentration of trace metals in clinical fluids of patients following ion release from different fixed metal orthodontic appliances.

Material	Source/ supplier	Body fluid/tissue analyzed	Metal ion(s) measured	Concentration ($\mu g\ L^{-1}$)		Detection mode	Sample pre-treatment	Refs.
				M	Control			
Stainless steel Arches	Discovery Unitek/3M Nitinol Remautium	Saliva	Cr Ni	2.6 18.5	2.2 11.9	AAS;L	1 mL of saliva sample was diluted with 10 mL of water	[179]
Stainless steel brackets	Dentaurum	Saliva	Cr Fe Ni	27 17 10	11 14 6	ICP-OES;L	10-15 mL of saliva sample was dried and mineralized using *aqua regia* in a closed system	[181]
Ni-Ti alloy Arches	Discovery Unitek/3M Nitinol Remautium	Mucosa cells	Co Cr Ni	0.84 4.24 21.74	0.44 3.46 12.26	GF AAS;L	Mucosa samples were diluted in water and acidified in HNO_3, kept at 60 C° for 10 min	[182]
Quad helix Head gear Fixed appliances (bands, brackets)	Not noted	Saliva	Cr Ni	78-108 65-85	61 55	GF AAS;L	0.5 mL of saliva sample was digested with 0.15 mL of conc. HCl. The saliva samples were centrifuged at 3000 x g for 2 min to settle particulate matter	[183]
Upper and lower fixed appliances (bands, brackets)	Ortho-cast Dentaurum Rematitan	Saliva	Cr Ni	29-800 7-332	54 53	AAS;L	0.5 mL of saliva samples were diluted with 10 mL of water	[184]
Bands Brackets Wires	Unitek Ormco	Saliva Serum	Cr Ni Cr Ni	0.53-1.53 4.12-11.53 6.16-10.98 7.87-10.27	0.76 4.45 6.21 8.36	GF AAS;L	Serum was prepared by centrifuging the blood samples at 3000 rpm for 10 min	[185]
Arches Bands Brackets	AISI Ormco Tru-Chrome SS	Mucosa cells	Co Ni	0.568 2.521	0.202 0.725	ICP-MS;L	1 mL of cell suspension from buccal mucosa cells was treated with HNO_3 (2 mL, 0.5%) and then diluted with water	[186]

Material	Brand	Sample	Element	Range	LOD	Technique	Sample preparation	Ref
Archwires	Elgiloy							
Arches Brackets Bands Archwires Wires	Unitek/3M Nitinol Ormco	Saliva plaque	Ni	0.005-25.25 µg g⁻¹	0.004 µg g⁻¹	ET AAS;L	Saliva samples were diluted with water; 0.7 g of the saliva was acidified in HNO₃	[187]
Dental implants (Not stated)	Not stated	Whole blood urine	Co Cr Mo Ni V Ti	0.073-0.402 0.094-0.528 0.309-0.790 0.451-0.957 0.053-0.068 0.316-1.770	Not reported	(HR/ORS)-ICP-MS;L	Blood samples were microwave-assisted, closed vessel digested with HNO₃/H₂O₂ and finally diluted tenfold with water. Urine samples were diluted tenfold with water	[188]
Brackets Brands Archwires Arch	Ormco Euro Ni-Ti	Saliva	Ni	56-78	34	ICP-MS;L	Saliva samples were dried, then digested with 0.2 mL *aqua regia*; solution obtained was diluted to a volume of 4 mL with water	[190]
Bonded brackets	3M/Unitek AISI Dentaurum	Saliva	Cr Fe Ni	0.29-1.72 28.31-103.58 1.69-16.01	Not reported	ETAAS;L	Saliva collection was prepared by drying in an electric oven at 150 °C for 15 min	[191]
Gold crowns	Conventional laboratory technique	Saliva	Ag Au Cu Pd Zn	0.2-46.7 15-38 8.3-63.5 13-49 695-1048	0.2 3.8 8.3 49 695	ICP-MS;L	*ca.* 2 mL of whole saliva sample was digested by adding 20 µL of conc. HNO₃; then, 10 mL of 1% HNO₃ was added to dilute the saliva	[192]
Inlays bridge	Degunorm Degulor	Urine	Au Pd Pt	19.9-187.7 20.2-143.2 10.5-59.6	6.1-184.3 12.4-121.4 1.0-7.4 ngL⁻¹		For stabilization purposes the samples were acidified with conc. HNO₃; then 5 mL aliquots of the acidified urine samples were mixed with 200 µL of H₂O₂ and UV irradiation (UV photolysis) for 15 min	[193]
Ti miniplates (ASTM F67-89)	Straumann Del-Tex	Soft tissues	Ti	0.09-2.33	Not reported	ICP-OES;L	Dry tissue samples were digested in a Teflon vessel with 65% HNO₃ in a microwave digestion system	[195]

L=liquid

Table 4. Selected operating parameters for trace metal determination in body fluids with detection by analytical atomic spectrometric techniques.

Orthodontic appliance	Spectrometer	Atomization/ excitation/ ionization source	Sample amount (mL)	Element (nm)/isotope	LOD[a] ng L^{-1}	Internal Quality Control/statistical analysis/CRMs/SRMs (or other validation)	Refs.
Arches	SpectrAA-220 Varian	Not noted[b]	Saliva, 10	Cr Ni	1000 1000	Mann–Whitney U-test, Wilcoxon W Statistical analysis $P<0.05$	[179]
Brackets	Optima 3000 Perkin-Elmer	ICP	Saliva, 25	Cr Fe Ni	1000 1000 1000	Calibration standards were formulated to be matrix-matched to the saliva-contained samples Two-way ANOVA $P=0.05$	[181]
Arches	SpectrAA-220 Varian	GF AAS	Mucosa cells	Co Cr Ni	1000 1000 1000	Student's t-test $P=0.05$	[182]
Quad helix Head gear Fixed appliance	ET-372 Perkin-Elmer	HGA 76B Graphite furnace	Saliva, 0.5	Cr 357.9 Ni 232.0	Not determined	Standard addition curve method Wilcoxon test	[183]
Upper and lower fixed appliances	Model 2380 Perkin-Elmer	Not noted[b]	Saliva, 10	Cr,Ni	Not determined	One-way ANOVA	[184]
Bands Brackets Wires	ATI 929 Unicam	Unicam 90 Graphite furnace	Saliva Serum	Cr,Ni	Not determined	Mann–Whitney U-test (SPSS statistic program). Synthetic saliva was used for the calibration curve	[185]
Arches Brands Brackets Archwires	HP 4500-200 Hewlett-Packard	ICP	Mucosa cells, 1	^{59}Co, ^{62}Ni	Not determined	Internal standard ^{103}Rh (0.25 µgL^{-1}). Mann–Whitney U-test from the Prism 3.0 Student's t-test	[186]

Sample	Instrument	Technique	Sample, amount	Element	LOD[a]	Method / Notes	Ref.
Brackets Arches Bands Archwires Wires	Zeeman 4100-ZL Perkin-Elmer	THGA Graphite furnace	Saliva, 2 Plaque, 2 µg	Ni	0.001-0.002 µgg^{-1}	Standard addition technique SPSS v.11.5 statistical package P<0.05	[187]
Dental implants	MAT Element Finnigen MAT Agilent 7500c Agilent Technologies	ICP	Whole blood, 10 Urine	^{59}Co ^{52}Cr ^{95}Mo ^{69}Ni ^{51}V ^{97}Ti	7-40 6-450 215-260 30-130 3-350 70-120	Internal standards: ^{71}Ga, ^{89}Y (10 µgL^{-1} each). Seronorm Trace Elements Whole Blood (Level 2, MR 9067) Seronorm Trace Elements Urine (Level 2, Ref. 201205)	[188]
Brackets Bands Arch Archwires	ELAN 5000 Perkin-Elmer	ICP	Saliva, 1	Ni	100	Wilcoxon signed rank test	[190]
Bonded Brackets	Analyst 800 Perkin-Elmer	GF AAS	Saliva	Cr Fe Ni	100 1000 1000	Standard calibration curve	[191]
Gold crowns	Not noted	ICP	Saliva, 10	Ag Au Cu Pd Zn	1000 1000 1000 1000 1000	Standard calibration curie	[192]
High-gold dental alloy	ELEMENT Finnigan MAT	ICP	Urine, 5	^{197}Au ^{106}Pd ^{195}Pt	0.2 0.2 0.2	Calibration was performed by the standard addition procedure	[193]
Ti miniplates	JY 24 Jobin Yvon-Horiba	ICP	Soft tissue, 13.4 mg	Ti 334.941	400	CRM Human Hair Powder (GBW07601) Two-way ANOVA	[195]

[a] LOD:
[b] Probably FAAS: flame atomic absorption spectrometry

Table 5. Selected operating parameters for trace metal determination in body fluids with detection by electrothermal atomic absorption spectrometry.

Prostheses	Spectrometer	Atomizer	Sample amount (mL)	Element (nm)	LOD[a] (μg L^{-1})	Internal quality control/statistical analysis/CRMs (or other validation)	Refs.
Knee	Zeeman 5100 Perkin-Elmer	HGA 600 graphite furnace	Serum	Ti	2.11	Aqueous standards were used for calibration	[42]
Knee	Zeeman Z-8270 Hitachi Ltd.	Graphite furnace	Blood, 7.5	Co Cr Mo Ni	0.5 0.5 0.5 0.5	Control material Seronorm Trace Elements Serum (SERO AS) Calibration was performed by the standard addition method	[43]
Knee	Not stated	Graphite furnace	Serum	Co, Cr, Mo	Not determined	Mann-Whitney U-test; P<0.05	[44]
Knee	Zeeman Z-8200 Hitachi Ltd.	Graphite furnace	Urine, 10 Blood, 10	Co Cr Ti	1 0.4 5	The concentrations were calculated using the standard addition method	[47]
Knee	High-resolution continuum source atomic absorption spectrometer Analytik Jena AG	ContrAA 700 Graphite furnace	Urine, 0.5	Mo 319.397 Ti 319.200	1.5 6.5	Matrix-matched standards were used for calibration. Urine reference materials: Seronorm Trace Element Urine and Clincheck Urine Control Level II	[48]
Hip	Zeeman 5100-ZL Perkin-Elmer	Graphite furnace	Blood Urine	Co Cr	0.1-0.4 0.3	Standard calibration technique Student's t-test	[53]
Hip	Zeeman 4100-ZL Perkin-Elmer	THGT Graphite tube	Blood, 20 μL	Cr 357.9 Co	0.11	Whole blood reference material IAEA-A-13 Prism software, P<0.05	[54]
Hip	Zeeman 4100-ZL Perkin-Elmer	Graphite furnace	Blood Urine	Co Cr Ni Mo	0.20 0.10 0.20 0.50	Certified Standards Trace Elements (NYCOMED) ANOVA	[55]

					3 nmolL^{-1}		
Hip	Zeeman SpectrAA800 Varian	Graphite furnace	Whole blood, 7	Cr		The determination was controlled with commercial-quality control materials	[68]
Hip	Solaar 939 QZ Unicam	Graphite furnace	Blood, 15 µL Serum	Co 240.7 Cr 357.9 Mo313.3	0.08 0.06 0.27	SRM 1598 NIST Human Serum	[93]
Hip	Solaar 939 QZ Unicam	Graphite furnace	Serum, 15 µL	Al Co Cr Ti	1.34 0.08 0.06 2.91	SRM 1598 NIST Human Serum Standard addition method	[95]
Hip	Zeeman 5100 Perkin-Elmer	HGA-600 Graphite furnace	Serum Urine	Co Cr	0.3 0.03	Spearman's rank-order correlation test	[100]
Hip	Zeeman 5100-ZL Perkin-Elmer	Graphite furnace	Serum	Co 242.5	0.3	Wilcoxon rank-sum test; P<0.05	[101]
Hip	Zeeman 5100	HGA-600 Graphite furnace	Serum	Co Cr Ti	0.3 0.03 2.22	Mann-Whitney test Wilcoxon rank sum tests Matrix-matched calibration curve	[115]
Hip	Zeeman Z-8200 Hitachi Ltd.	Graphite furnace	Serum	Co Cr Mo	0.5 0.5 0.5	Control materials Seronorm Trace Elements Serum (SERO AS) Standard addition method	[117]
Hip	Zeeman 5100-ZL Perkin-Elmer	Graphite furnace	Serum	Al 309.3 Co 242.5	0.5 0.3	P $<$ 0.05	[118]
Hip	Zeeman 5100 SIMAA 6100 Perkin-Elmer	Graphite furnace	Serum	Co Cr	0.3 0.3	Seronorm levels I and II (SERO AS) ANOVA	[125]
Hip	JY24 Jobin Yvon	ICP		Ti	1.4		
Hip	Solaar 939 QZ Unicam	Graphite furnace	Serum	Al Co Cr Mo	1.36 0.08 0.06 0.27	Stat View 5.01 for Windows	[137]

	Instrument	Technique	Sample	Element	Ti	LOD[a]	Reference material	
					2.91			
Hip	Zeeman 5100 Perkin-Elmer	HGA-600 Graphite furnace	Serum	Co		0.3	Seronorm Trace Elements Serum (SERO AS)	[139]
				Cr		0.03		
Hip	Solaar 939 QZ Unicam	Graphite furnace	Serum, 15 µL	Co		0.08	SRM 1598 NIST Human Serum	[141]
				Cr		0.06	Standard addition method	
				Mo		0.83		
				Ni		0.10		

[a] LOD, Limit of detection

Table 6. Selected operating parameters for trace metal determination in body fluids with detection by mass spectrometry.

Prostheses	Mass spectrometer	Internal standards	Sample amount (mL)	Element/isotope	LOD[a] (ng L^{-1})	Internal quality control/statistical analysis/CRMs/SRMs (or other validation)	Refs.
Hip, knee	Element 2 Finnigan MAT	^{71}Ga, ^{89}Y	Blood, 5 Urine	^{47}Ti ^{51}V ^{52}Cr ^{59}Co ^{60}Ni ^{98}Mo	66.4 1.5 4.4 0.8 28.6 0.8	Seronorm Trace Elements Urine (Level 2, Ref. 201205), Seronorm Trace Elements Whole Blood (Level 2, MR 9067)	[46]
Hip	Plasmatrace II Thermo Scientific	In	Whole Blood Urine	Co Cr Ti V	0.07 1.0 0.1 0.4-0.04	Calibration was performed using a method of standard additions technique	[57]
Hip	DRC Elan Perkin-Elmer		Blood, 10	Co, Cr	Not determined	Student's t-test $P<0.05$	[58]
Hip	SCIEX Elan 6100 Perkin-Elmer		Whole Blood, 1	Co, Cr, Mo	10-30	Reference material: Seronorm Trace Elements Whole Blood (Level 2)	[59]
Hip	Finnigan MAT Thermo Electron Corporation		Whole Blood, 5-6 Urine	Co,Cr	20-60	t-test, $P<0.05$	[60,63]
Hip	Varian Ultramass		Whole Blood, 7	Co	2 nmolL^{-1}	The determination was controlled with commercial-quality control materials	[68]
Hip	7500c Agilent		Whole Blood Plasma Urine	Co,Cr,Ni	Not determined	ANOVA	[73]

Joint	Instrument	Internal standard	Sample, volume	Element/isotope	LOD	Method / Reference material	Ref.
Hip	DRC Elan 6100 Perkin-Elmer		Whole Blood	Co, Cr	Not determined	Seronorm Trace Elements Whole Blood	[74]
Hip, Knee	Element Finnigan MAT	^{103}Rh	Whole Blood, 1	^{52}Cr, ^{59}Co, ^{60}Ni, ^{95}Mo	60, 60, 300, 60	Seronorm Trace Elements Whole Blood, Level 2 (Nycomed AS) External calibration method	[82]
Hip	SCIEX Elan 6100 Perkin-Elmer	Ga, Rh	Serum, 10	Co, Cr	Not determined	Seronorm Trace Elements Whole Blood (Nycomed)	[104]
Hip	Element I Finnigan MAT	^{115}In	Serum, 5 / Urine	^{59}Co, ^{52}Cr, ^{55}Mn, ^{100}Mo, ^{60}Ni	10-20, 20, 10-30, 40-50, 10-30	Mann-Whitney U-test	[114]
Hip	Elan 6100 SCIEX Perkin-Elmer	Ga, Rh	Serum	Co, Cr	2 nmolL^{-1}, 2 nmolL^{-1}	The Kruskal-Walli's test	[119]
Hip	Elan DRC II Perkin-Elmer		Blood, 5 / Serum	Co, Cr	500, 500	Mann-Whitney U-test	[122]
Hip	Element II Thermo Fisher Scientific		Blood, 5 / Urine	^{59}Co, ^{52}Cr, ^{55}Mn, ^{95}Mo, ^{47}Ti	1-200, 1-200, 1-200, 1-200, 1-200	Seronorm Trace Elements Urine (Level 1, Ref. 201205), Seronorm Trace Elements Whole Blood (Level 1, Ref 9067)	[130]
Hip	Not noted		Blood / Serum	Co, Cr	300, 30	Mann-Whitney testing	[131]
Nails	ELEMENT II Thermo Fisher Scientific		Serum, 0.6	^{47}Ti	0.05	Seronorm Trace Elements Serum (Level 1, Ref 201405, Nycomed AS)	[136]
Hip	ELEMENT 2 Thermo Fisher Scientific		Serum	Co, Cr	40, 15	Seronorm Trace Elements Serum (SERO AS)	[139]

					LOD[a]		
Hip	DRC Elan 6100 Perkin-Elmer SCIEX		Serum	Co Cr	0.2 0.2	Wilcoxon signed rank test	[147]
Hip	ELEMENT 2 Thermo Fisher Scientific		Serum	Co Cr	Not determined	Wilcoxon exact rank – sum test	[150]
Hip	HR-SF-ICP-MS Thermo Fisher Inc.		Serum Urine Hair	Co,Cr Co,Cr Co,Cr,Mo	Not determined	Wilcoxon's non-parametric test $P<0.05$	[160]
Hip	HR-SF-ICP-MS ELEMENT 2 Thermo Fisher Scientific	^{89}Y	Whole blood 5	Co Cr Ti	20 100 200	Reference samples ANOVA	[217]
Hip	Agilent 8800 ICP-QQQ-MS/MS	^{71}Ga	Serum	^{47}Ti	8	Seronorm Trace Elements Serum, Level 1 (Ref. 9067 – Sero)	[218]

[a] LOD, Limit of detection

Table 7. Clinical CRMs, SRMs or RMs provided by government, international agencies or the private sector with certified trace element concentrations.

Source/supplier	Material/product	Description	Elements certified (info. only)
NIST[a]	SRM 909b	Human Serum	Ca, Li, Mg, K, Na
	SRM 909c	Human Serum	Se, Na
	SRM 1950	Metabolites in Human Plasma	Cu, Se, Zn
	SRM 95c	Electrolytes in Frozen Human Serum	Ca, Li, Mg, K, Na
	SRM 2670a	Toxic Elements in Urine (Freeze-dried)	Sb, Ca, Co, Pb, Hg, Mn, Mo, Pt, Se, Tl (Al), (As), (Ba), (Be), (Cr), (Cu), (Mo), (Ni), (Sn), (W), (V)
	SRM 2672a	Mercury in Urine	Hg
	SRM 1577c	Bovine Liver	Ag, As, Ca, Cd, Co, Cr, Cu, Fe, K, Mg, Mn, Mo, Na, Ni, Pb, Se, Sr, V, Zn (Li)
IRMM[b]	ERM-CE 194/195/196	Lyophilized Bovine Blood	Pb, Cd
	BCR-634/635/636	Lyophilized Human Blood	Pb, Cd
	BCR-304	Lyophilized Human Serum	Ca, Li, Mg
	BCR-637/638/639	Human Serum	Al, Se, Zn
	BCR-185R	Bovine Liver	As, Cd, Cu, Mn, Pb, Se, Zn
	BCR-273/274	Single Cell Protein	As, Ca, Cd, Co, Cu, Fe, K, Mn, Pb, Se, Zn
IAEA[c]	RM 086	Human Hair	Hg, Fe, Zn
	RM A-13	Animal Blood	(Ca), (Cu), (Mg), (Mn), (Se), (Sc) Ca, Cu, Fe, K, Mg, Na, Ni, Pb, Se, Zn
NIES[d]	CRM No. 18	Human Urine	As, Se, Zn (Cu), (Pb)
	CRM No. 13	Human Hair	Hg, Cd, Cu, Pb, Sb, Se, Zn (Al), (Ag), (As), (Ba), (Ca), (Co), (Fe), (Mg), (Mn), (Na), (V)
LGC Standards[e]	BCR-636	Reconstituted Human Blood	Cd, Pb
SERO AS[f]	Seronorm L-1-2-3	Trace Elements Whole Blood	Al, Be, Cr, Mn, Ni, Sn, Sb, Bi, Co, Hg, Se, V, As, Cd, Cu, Pb, Mo, Tl, Zn
	Seronorm L-1-2	Trace Elements Serum	Al, Co, Au, Mn, Ni, Zn, Ca, Cu, Fe, Mg,

| Seronorm L-1-2 | Trace Elements Urine | | Se, Cr, Hg |
| VIRM[g] | Information Center Data from various databases for reference materials | | Al, Be, Cr, Hg, Te, V, Sb, Bi, Co, Pb, Ni, Tl, Zn, As, Cd, Mn, Se, Sn |

[a] National Institute of Standards and Technology (NIST, Gaithersburg, MD USA); http://www.nist.gov
[b] Institute for Reference Materials and Measurements (Geel, Belgium); http://www.irmm.jrc.be
[c] International Atomic Energy Agency (Vienna, Austria); http://www.iaea.org
[d] National Institute for Environmental Studies (NIES, Ibaraki, Japan); http://www.nies.go.jp
[e] LGC Standards; http://www.lgcstandards.com
[f] SERO AS (Seronorm, Billingstad, Norway); http://www.sero.no
[g] Virtual Institute for Reference Materials; http://www.VIRM.net

Printed by Books on Demand GmbH, Norderstedt / Germany